DR.-ING. J. WANKE – DIPL.-ING. M. HAVLÍČEK
unter Mitwirkung von Dr.-Ing. A. Warner

ENGLISCH
FÜR ELEKTROTECHNIKER

OSCAR BRANDSTETTER VERLAG KG · WIESBADEN

VDE-VERLAG GMBH · BERLIN

355/84

J. WANKE – M. HAVLÍČEK
with
collaboration of A. Warner

ENGLISH
FOR ELECTRICAL ENGINEERS

OSCAR BRANDSTETTER VERLAG KG · WIESBADEN

VDE-VERLAG GMBH · BERLIN

Gemeinschaftsausgabe
der OSCAR BRANDSTETTER VERLAG KG, Wiesbaden
und der VDE-VERLAG GmbH, Berlin

In diesem Buch werden, wie in allgemeinen Nachschlagewerken üblich, etwa bestehende Patente, Gebrauchsmuster oder Warenzeichen nicht erwähnt. Wenn ein solcher Hinweis fehlt, heißt das also nicht, däß eine Ware oder ein Warenname frei ist.

In this book as in general reference works existing patents, patterns, trade marks etc. are not mentioned. If such a reference is absent, it does not mean that a commodity or tradename is "free".

1. Auflage 1971
Copyright © 1971 by
OSCAR BRANDSTETTER VERLAG KG, WIESBADEN
Gesamtherstellung: Oscar Brandstetter Druckerei KG, Wiesbaden
ISBN 3 87097 052 9
Printed in Germany

Inhalt

Einleitung VII
Ausgewählte Grundbenennungen. X

Fundamentals of Electrical Engineering (Grundlagen der Elektrizitätslehre)
1. Basic Electricity (Grundzüge der Elektrizitätslehre) 1
2. Electrical and Related Quantities and Units (Elektrische und verwandte Größen und Einheiten) 10
3. Electrical Materials (Werkstoffe der Elektrotechnik) 21
4. Manufacture of Electrical Equipment (Herstellung elektrischer Geräte) . . . 28
5. Electrical Measurements (Elektrische Messungen) 36

Electric Power Engineering (Starkstromtechnik)
6. Electric Power Production (Erzeugung elektrischer Energie) 47
7. Transmission and Distribution of Electric Power (Übertragung und Verteilung elektrischer Energie) 54
8. Electrical Machinery (Elektrische Maschinen) 60
9. Electric Light and Heat (Elektrisches Licht und Wärme) 67
10. Electrochemistry and Electrometallurgy (Elektrochemie und Elektrometallurgie) 75

Electronic Engineering (Elektronik)
11. Semiconductor Devices and Integrated Circuits (Halbleiterbauelemente und integrierte Schaltungen) 81
12. Electron Tubes (Elektronenröhren) 93
13. Laser Engineering (Lasertechnik) 105

Control and Computer Engineering (Steuer-, Regel- und Computertechnik)
14. Pulse Technique (Impulstechnik) 111
15. Control & Automation (Regelung, Steuerung und Automatisierung) 116
16. Data Processing (Datenverarbeitung) 131

Communication Engineering (Nachrichtentechnik)
17. Principles of Electrical Communication (Grundlagen des elektrischen Nachrichtenwesens) 142
18. Wire Communications (Draht-Fernmeldewesen) 149
19. Sound Transmission and Recording (Schallübertragung und -Aufnahme) . . 160

Radio Engineering (Funktechnik)
20. Radio Transmitters (Funksender) 173
21. Radio Receivers (Funkempfänger) 183
22. Antennas and Transmission Lines (Antennen und Übertragungsleitungen) . 192
23. Radio Wave Propagation (Funkwellen-Ausbreitung) 200
24. Microwave and Space Communications (Mikrowellen- und Weltraum-Nachrichtentechnik) 207
25. Television (Fernsehen) 217
26. Radio Navigation and Radar (Funknavigation und Radar) 231

Einführung in das technische Englisch 244
Unterschiede zwischen der britischen und amerikanischen Fachsprache 247
Abkürzungen in der elektrotechnischen Fachsprache 250
Fremdsprachige Ausdrücke in der englischen Fachsprache 257
Englische Konferenzterminologie. 262
Akademische Titel und Berufsbezeichnungen im englischen Fachschrifttum. . . 266
Mathematik und mathematische Logik in der englischen Fachsprache 268

ANHANG: Über fachsprachliche Wendungen (Dr.-Ing. A. Warner) 276
Sachregister 281

Einleitung

Das vorliegende Handbuch stellt einen Versuch dar, elektrotechnisches Englisch dem Techniker, Ingenieur, Übersetzer und Dolmetscher, Dokumentalisten und Studenten zu vermitteln. In erster Linie handelt es sich um ein terminologisches Hilfsmittel, dessen Studium ein gewisses Minimum von allgemeinen englischen Sprachkenntnissen voraussetzt, vor allem was die Grammatik betrifft.

Den Kern der Übungstexte dieses Buches bilden mehr oder weniger zubereitete Originalaufsätze aus allen Gebieten der modernen Elektrotechnik. Einige Aufsätze, die durch eine Quellenangabe gekennzeichnet sind, wurden – geringfügig gekürzt – dem modernen angloamerikanischen Fachschrifttum entnommen, damit der Leser die ursprüngliche britische oder amerikanische Darstellungsweise und den Stil der Fachaufsätze mit allen ihren Vorzügen und Mängeln kennenlernt.

Da in den Aufsätzen bekannte Themen behandelt werden, ist dem Leser das Studium in der Weise erleichtert worden, daß sich ihm neue und unbekannte Ausdrücke und Wendungen nahezu von selbst aus dem Zusammenhang erklären.

Außer der Hauptaufgabe, ins moderne Englisch der Elektrotechnik und der Elektronik einzuführen, wie man es in Großbritannien, in den USA und bei internationalen Verhandlungen benutzt, stellt sich das Buch noch ein weiteres Ziel: es weist auf Unterschiede der Terminologie in den verschiedenen Fach- und Sprachgebieten hin. Dabei wird selbstverständlich die nationale – d. h. britische und amerikanische – sowie die internationale Normung berücksichtigt.

Die Arbeit an solch einem Werk muß immer einen Kompromiß zwischen zwei Gesichtspunkten darstellen: einerseits ist es das Bestreben, soviel Stoff wie möglich zu erfassen, andererseits besteht die Notwendigkeit, einen gewissen Umfang des Werkes nicht zu überschreiten. Das Buch ist also kein Fachwörterbuch! Im Gegenteil, sein zweckmäßiger Gebrauch setzt die gleichzeitige Benutzung eines guten Wörterbuches (z. B. ERNST 1971) und/oder Lexikons voraus, z. B. New Oxford Dictionary (britisch) oder Webster's Collegiate Dictionary (amerikanisch).

Beim Lernen einer fremden Fachsprache wird unumgänglich auch die eigene gelernt. Der Student muß sich nicht nur englische, sondern auch „richtige" (d. h. vereinbarte oder in Normen empfohlene) deutsche Benennungen einprägen. Deshalb überschreiten oft in den „Fachsprachlichen Anmerkungen" die Erläuterungen die Grenze des Englischen und reichen tiefer als üblich ins Deutsche hinein. So werden bisweilen auch Hinweise zur Benutzung von deutschen Benennungen gegeben, insbesondere dort, wo die deutsche Fachsprache zwei Begriffe kennt, für die es im Englischen nur eine einzige Benennung gibt (z. B. Geräusch und Rauschen für *noise*).

Das Einbeziehen der Fachbenennungen in die Texte stützt sich vor allem auf die Häufigkeit, mit der die jeweilige Benennung im Fachschrifttum vorkommt sowie auf die Bedeutung, die ihr beigemessen wird. Natürlich unterliegen beide Kriterien einer mehr oder weniger subjektiven Beurteilung. Man muß sich auch darüber klar sein, daß neue und interessante Gebiete der Wissenschaft und Technik nicht immer sogleich über einen umfangreichen Wortschatz verfügen.

Einige Benennungen in den Wortverzeichnissen und den „Ausgewählten verwandten Benennungen" sind mit einem Sternchen gekennzeichnet und am Schluß des Kapitels in sogenannten „Fachsprachlichen Anmerkungen" näher erläutert. Unter den Stichwörtern dieser Anmerkungen werden u. a. auch synonyme, semantische, morphologische oder etymologisch usw. verwandte Benennungen angeführt.

Es wird dabei auch auf feine semantische Unterschiede zwischen Synonymen hingewiesen, und es werden Nuancen hervorgehoben, die in den meisten Wörterbüchern gewöhnlich nicht herausgearbeitet werden. Auch Ungenauigkeiten, Unregelmäßigkeiten und der nicht konsequente Gebrauch vieler Benennungen werden in den „Anmerkungen" erörtert.

Ein Verzeichnis der erläuterten Benennungen mit Angabe des entsprechenden Kapitels ist am Ende des Handbuches angefügt (Sachregister).

Einige Benennungen in den Wortverzeichnissen und fachsprachlichen Anmerkungen sind durch Länderzeichen ergänzt. Diese kennzeichnen den räumlichen Gültigkeitsbereich oder die regionalen Besonderheiten der jeweiligen Benennung:

/A/	Österreich
/CH/	Schweiz
/D/	Deutschland
/GB/	Großbritannien
/USA/	Vereinigte Staaten von Amerika

Viel öfters sind den Benennungen sog. Autoritätszeichen angehängt, deren Zweck es ist, autoritative Quellen anzugeben, welchen die jeweilige Benennung entnommen worden ist. Als solche gelten Unterlagen, besonders Wörterbücher, deren Inhalt von nationalen oder internationalen fachlichen und überfachlichen Organisationen vereinbart worden ist.

Autoritätszeichen werden im vorliegenden Werk besonders bei Benennungen angeführt, die in den Schriftstücken dieser Organisationen definiert werden, sonst aber im Fachschrifttum nicht immer konsequent benutzt werden oder Bedenken erregen könnten.

Es werden folgende Autoritätszeichen benutzt:

Autoritätszeichen	Bedeutung
AEF	Ausschuß für Einheiten und Formelgrößen im DNA
ANSI	American National Standards Institute (vormals: American Standards Association – ASA)
BS	Normenzeichen des British Standards Institution (BSI)
CEE	International Commission on Rules for the Approval of Electrical Equipment
CGPM	Conférence Générale des Poids et Mesures
CIE	Commission Internationale d'Eclairage
CISPR	International Special Committee on Radio Interference
DBP	Deutsche Bundespost
DGON	Deutsche Gesellschaft für Ortung und Navigation
DIN	Deutsche Industrie-Normen (Verbandszeichen des DNA)
DNA	Deutscher Normenausschuß
EBU	European Broadcasting Union
EOQC	European Organization for Quality Control
GAMM	Gesellschaft für Angewandte Mathematik und Mechanik
IBM	International Business Machines Corp., New York; Internationale Büro-Maschinen GmbH., Sindelfingen
ICAO	International Civil Aviation Organization
IEC	International Electrotechnical Commission
IEEE	Institution of Electrical and Electronic Engineers (USA)
ISO	International Organization for Standardization
ITU	International Telecommunication Union
NARTB	National Association of Radio and Television Broadcasters (USA)

NRC	National Research Council (USA)
NTG	Nachrichtentechnische Gesellschaft im VDE
ÖVE	Österreichischer Verband für Elektrotechnik
p...	Entwurf (z. B. pDIN, pIEC)
UIP	Reihentitel der International Union for Pure and Applied Physics (UIPAP)
VDE	Verband Deutscher Elektrotechniker
VDI	Verein Deutscher Ingenieure
VOFu	Vollzugsordnungen für den Funkdienst

* * *

Wir möchten abschließend nicht versäumen, Herrn Dr.-Ing. Alfred WARNER (Darmstadt) für die tatkräftige Mitarbeit und für wertvolle Anregungen zum Inhalt und zur Gesamtanlage des Buches unseren herzlichsten Dank zu sagen. Es war seine Idee, die Wörterverzeichnisse mit Autoritätszeichen zu versehen (der größte Teil der Autoritätszeichen ist dann von ihm nachgetragen worden). Auch der von Dr.-Ing. WARNER verfaßte Abschnitt „Fachsprachliche Wendungen" im Anhang stellt einen wertvollen Impuls für die Vertiefung unserer Kenntnisse der Fachsprache dar.
Bester Dank gebührt auch dem Verlag und der Druckerei, die mit uns bei der Disposition, bei der konkreten Gestaltung und dann bei der Drucklegung des Manuskripts vorbildlich zusammengearbeitet haben.
Herbst 1971

Die Autoren

Ausgewählte Grundbenennungen

Die vorliegende Zusammenstellung enthält über hundert englische Benennungen, die in den durchgearbeiteten englischen Texten besonders häufig vorkamen oder die im elektrotechnischen Fachschrifttum ohne Ansehen des jeweiligen Spezialgebietes immer wieder gebraucht werden, so daß sie als unentbehrlich angesehen werden können. Deshalb ist es ratsam, das Studium dieses Buches nicht fortzusetzen, ohne die Kenntnis dieser Grundbenennungen überprüft zu haben.

Allerdings wird hier nur die allgemeine oder die Grundbedeutung jedes englischen Wortes angeführt, während alle abgeleiteten oder Sonderbedeutungen (auch aus anderen als elektrotechnischen Gebieten) von Fall zu Fall in den Wortverzeichnissen zu den Übungstexten zu finden sind. So wird hier z. B. für *shunt* nur „Nebenschluß" angegeben, nicht „Nebenschlußwiderstand" oder „Nebenschlußkreis", für *load* nur „Last", nicht etwa „Belastung", „Verbraucher", „Beanspruchung", „Bespulung" usw.

Wenn auch die hier angeführten Benennungen zum Grundwortschatz des Elektrotechnikers gehören, so soll damit nicht gesagt werden, das ihre Übersetzung problemlos ist. Erläuterungen zu den für das Übersetzen schwierigen Fällen können über das Register zu den „Fachsprachlichen Anmerkungen" aufgesucht werden.

Um Platz zu sparen, wurden hier (wie auch in die übrigen Teile des Buches) nicht solche englischen Termini einbezogen, die im Deutschen gleichlautend oder sehr ähnlich sind und eindeutig übertragen werden können (z. B. *anode*, *impedance*, *motor*, usw.).

alternating current; a. c.	Wechselstrom *m*
ammeter	Strommesser *m*
amplification	Verstärkung *f*
amplifier	Verstärker *m*
arc	Bogen *m*
attenuation	Dämpfung *f*
audio	Hör-; akustisch
bridge	Brücke *f*
cable	Kabel *n*
capacitance	Kapazität *f*
capacitor	Kondensator *m*
cell	Zelle *f*
channel	Kanal *m*
charge	Ladung *f*
choke	Drossel *f*
circuit	Kreis *m*
coil	Spule *f*
communication	Verbindung *f*
computer	Rechenanlage *f*; Rechner *m*
conductance	Leitwert *m*
conduction	Leitung *f* (Vorgang)
conductivity	Leitfähigkeit *f*
conductor	Leiter *m*
connection	Verbindung *f*
contactor	Schütz *n*
control	Regelung *f* und Steuerung *f*
core	Kern *m*
coupling	Kopplung *f*
current	Strom *m*

cycle	Periode *f*; Schwingung *f*
damping	Dämpfung *f* (von Schwingungen)
direct current; d.c.	Gleichstrom *m*
discharge	Entladung *f*
distortion	Verzerrung *f*; Verzeichnung *f*
drop	Abfall *m*
earth /GB/	Erde *f*
eddy currents	Wirbelströme *m pl*
efficiency	Wirkungsgrad *m*
e.m.f.; electromotive force	EMK; elektromotorische Kraft *f*
to feed	speisen
feedback	Rückkopplung *f*
field	Feld *n*
flux	Fluß *m*
fuse	Sicherung *f*
gain	Gewinn *m*
ground /USA/	Erde *f*
inductance	Induktivität *f*
inductor	Spule *f*
input	Eingang *m*
insulation	Isolation *f*
insulator	Isolator *m*
lag	Nacheilung *f*
lead [li:d]	Leiter *m*; Voreilung *f*
line	Leitung *f*
load	Last *f*
loop	Schleife *f*
loss	Verlust *m*
to match	anpassen
measurement	Messung *f*
meter	Meßgerät *n*
mutual inductance	Gegeninduktivität *f*
network	Netzwerk *n*
noise	Geräusch *n*; Rauschen *n*
output	Ausgang *m*; Ausgabe *f*
particle	Teilchen *n*
peak	Spitze *f*
pole	Pol *m*; Mast *m*
power	Leistung *f*
power factor	Leistungsfaktor *m*; Wirkfaktor *m*
primary	primär; Primärwicklung *f*
propagation	Ausbreitung *f*
pulse	Impuls *m*
radio	Funk-
ratio	Verhältnis *n*
reactor	Reaktor *m*; Drossel *f*
receiver	Empfänger *m*
reception	Empfang *m*
rectifier	Gleichrichter *m*
relay	Relais *n*
resistance	Widerstand *m* (Eigenschaft)
resistivity	spezifischer Wiederstand *m*

resistor	Widerstand *m* (Bauelement)
r.m.s. value; root-mean-square value	Effektivwert *m*
secondary	sekundär; Sekundärwicklung *f*
self-inductance	Selbstinduktivität *f*
semiconductor	Halbleiter *m*
shield; shielding	Abschirmung *f*
short circuit	Kurzschluß *m*
shunt	Nebenschluß *m*
solid state	Festkörper *m*
switch	Schalter *m*
telecommunications	Fernmeldewesen *n*; Nachrichtentechnik *f*
terminal	Klemme *f*
transmission	Sendung *f*; Übertragung *f*
transmitter	Sender *m*
tube /USA/	Elektronenröhre *f*
turn	Windung *f*
valve /GB/	Elektronenröhre *f*
voltage	Spannung *f*
wave	Welle *f*
winding	Wicklung *f*
wire	Draht *m*
wireless	drahtlos

Fundamentals of Electrical Engineering
(Grundlagen der Elektrizitätslehre)

1. Basic Electricity
(Grundzüge der Elektrizitätslehre)

1.1. Concepts of the Nature of Electricity
(Über das Wesen der Elektrizität)

Electricity comprises physical *phenomena* involving *electric charges* and their effects when *at rest* and when *in motion*. Electricity is manifested as a *force of attraction* when two *oppositely charged* bodies are brought close to one another. The elementary electric charges are possessed by electrons and protons, the charge of an electron being equal in *magnitude* to that of a proton, but *electrically opposite*. The electron's charge is arbitrarily termed negative, and that of the proton, positive.

If two metallic bodies, one with a positive charge and the other with a negative charge, are joined by a metal wire, *electric current* caused by the *electromotive force* (abbreviated e.m.f.) will flow from one to the other. The metal is, therefore, a good *conductor* of electricity, or it offers a low *resistance* to the passage of electricity through it.

In contradistinction, glass is a very poor conductor and is called a *nonconductor* or *insulator*. A few substances, including *silicon*, selenium and germanium, have electrical *conductivities* about midway between good conductors and good insulators and are known as *semiconductors*.

An *electric field* (or *electrostatic field*, also termed *dielectric field*) exists in a region if an electric charge at rest in the region *experiences* a force of electrical origin. The electric field is conveniently represented by *electric lines of force* or *lines of electric flux*. Another useful term applied for describing the properties of electric fields is the *electric field strength* (or *electric field intensity*).

A changing electric field always produces a *magnetic field* and the resulting *interaction* of electric and magnetic forces gives rise to a region in space known as an electromagnetic field.

That branch of *electrical science* which treats the phenomena associated with electricity at rest is termed *electrostatics* while *electrokinetics* deals with the laws governing electricity in motion.

at rest	im Ruhezustand
conductivity* IEC, UIP	Leitfähigkeit *f* DIN, UIP
conductor IEC	Leiter *m* DIN, VDE, IEC
electrically opposite	elektrisch entgegengesetzt
electrical* science	Elektrizitätslehre *f*
electric* charge IEC	elektrische Ladung *f* DIN
electric current IEC	elektrischer Strom *m* DIN
electric field IEC; electrostatic field; dielectric field	elektrisches Feld *n* DIN
electric field strength UIP; electric field intensity	elektrische Feldstärke *f* DIN, UIP
electric line of force	elektrische Feldlinie *f*
electrokinetics IEC	Elektrokinetik *f*; Lehre *f* von der elektrischen Strömung IEC
electromotive force UIP; e.m.f.	elektromotorische Kraft *f* DIN, UIP

electrostatics	Elektrostatik *f*
to experience	unterliegen
force of attraction	Anziehungskraft *f*
in contradistinction	demgegenüber
in motion	in Bewegung
insulator IEC	(hier:) Isolierstoff *m*
interaction	Wechselwirkung *f*
line of electric flux	elektrische Feldlinie *f*
magnetic field	magnetisches Feld *n* DIN
magnitude	(hier:) Betrag *m*
nonconductor	Nichtleiter *m*
oppositely charged	entgegengesetzt geladen
phenomenon (pl. phenomena)	Erscheinung *f*
resistance* IEC, UIP	Widerstand *m* DIN, IEC, UIP
semiconductor IEC	Halbleiter *m* DIN, IEC
silicon	Silizium *n*

A selection of related terms — **Ausgewählte verwandte Benennungen**

circulation of a vector	Hüllenintegral *n*; Randintegral *n* DIN
curl field IEC	Wirbelfeld *n*
current density IEC, UIP	elektrische Stromdichte *f* DIN, UIP
density of electric charge	Ladungsdichte *f*
linear ~ ~ ~ ~	lineare ~
surface ~ ~ ~ ~ ~ ; surface charge density UIP	Flächen~ DIN, UIP
volume ~ ~ ~ ~ ~ ; charge density UIP	Raum~ DIN, UIP
dielectric polarization UIP	elektrische Polarisation *f* DIN, UIP; dielektrische Polarisation *f*
electric dipole moment UIP	elektrisches Dipolmoment *n* DIN, UIP
electric induction density ISO; electric flux density IEC; electric displacement IEC, ISO, UIP	elektrische Flußdichte *f* DIN; elektrische Verschiebungsdichte *f* DIN, UIP; elektrische Verschiebung *f* DIN, UIP
electric potential UIP	elektrisches Potential *n* UIP
electric susceptibility UIP	elektrische Suszeptibilität *f* DIN, UIP
flux of displacement ISO; electric displacement flux ISO; electric flux of induction ISO; electric flux IEC, UIP	elektrischer Fluß *m* DIN; Verschiebungsfluß *m* DIN; elektrischer Induktionsfluß *m*
irrotational field IEC	wirbelfreies Feld *n*
potential difference UIP; tension UIP; voltage IEC	elektrische Spannung *f* DIN, IEC, UIP; Potentialdifferenz *f* UIP
potential gradient	Potential-Gradient *m*
rotation; curl	Rotation *f*; Wirbeldichte *f*
solenoid field IEC	quellenfreies Feld *n*

1.2. Fundamentals of Magnetism
(Grundlagen des Magnetismus)

Magnetism comprises physical phenomena involving magnetic fields and their effects upon materials. Magnetic phenomena are associated with *moving charges*; electrons, considered as *particles*, are assumed to possess an *axial spin*, which gives the effect of a *minute current-turn* or of a small *permanent magnet*, called a Bohr magneton. The gyroscopic effect of an electron spin develops a *precession* when a magnetic field is applied.

The space surrounding a *magnetic pole* or a current of electricity in which magnetic forces act is called a magnetic field. The direction of the magnetic force at any point in the field is the direction in which a *unit north pole* placed at the point *tends* to move, and its magnitude is the force in newtons *exerted* on the unit pole. A curve drawn in the direction of the magnetic force is called a *magnetic line of force*. The magnetic field is conveniently represented by *lines of magnetic flux* (or *magnetic induction* or *magnetic flux density*).

axial spin	axialer Spin *m*
current-turn	Stromwindung *f*
to exert	ausüben
line of magnetic flux	magnetische Feldlinie *f*
magnetic induction UIP; magnetic flux density UIP	magnetische Induktion *f* DIN; magnetische Flußdichte *f* DIN
magnetic line of force	magnetische Feldlinie *f*
magnetic pole	magnetischer Pol *m*
minute	winzig
moving charge	Ladung *f* in Bewegung
particle	Teilchen *n*
permanent magnet	Dauermagnet *m*
precession	Präzession *f*
to tend	neigen
unit north pole	Einheits-Nordpol *m*

A selection of related terms — **Ausgewählte verwandte Benennungen**

coercive force	Koerzitivfeldstärke *f* DIN
flux of the magnetic induction ISO; magnetic flux IEC, UIP	magnetischer Induktionsfluß *m* DIN; magnetischer Fluß *m* DIN, UIP
hysteresis	Hysterese *f*
magnetic creep	magnetische Nachwirkung *f*
magnetic dipole moment	Coulombsches magnetisches Moment *n* DIN
magnetic field intensity ISO; magnetic field strength IEC, UIP, BS; magnetising force BS	magnetische Feldstärke *f* DIN, UIP; magnetische Erregung *f*
magnetic permeance	magnetischer Leitwert *m*
magnetic polarization ISO, UIP; intrinsic magnetic flux density IEC	magnetische Polarisation *f* DIN, UIP
magnetic potential difference UIP	magnetische Spannung *f* DIN, UIP
magnetization IEC, UIP; intensity of magnetization ISO	Magnetisierung *f* DIN, UIP
magnetomotive force UIP; m.m.f.	magnetomotorische Kraft *f* UIP
permeability UIP	Permeabilität *f* DIN, UIP
permeance	magnetischer Leitwert *m*; Permeanz *f*
relative permeability	Permeabilitätszahl *f* DIN, UIP
reluctance	magnetischer Widerstand *m*; Reluktanz *f*
residual flux density	Remanenz *f* DIN

1.3. Electric Currents and Circuits
   (Elektrische Ströme und Stromkreise)

An electric current may be continuous (the *direct current*, abbreviated *d.c.*), or it may flow first in one direction through the circuit and then in the other (the *alternating current*, abbreviated *a.c.*), or it may consist of *temporary surges* (*transients*).

The *reversals* of alternating currents may occur at any rate from a few per second up to several billion per second. The typical *waveform* of a *pure sinusoidal* alternating current is shown in Fig. 1–1. The number of *cycles* in one second is called the *frequency*, the time for one complete cycle is the *period*.

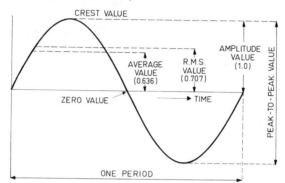

Fig. 1–1. Waveform of one cycle of a sine wave.
(Form einer Periode eines Sinusvorganges.)

amplitude value	Amplitude; Scheitelwert (bei einer Sinus-Schwingung)
average value; mean value	Mittelwert; Gleichwert
crest value	Scheitelwert
peak-to-peak value	(hier:) Schwingungsbreite
period	Periodendauer
r.m.s. value; root mean square value; effective value	Effektivwert
zero value	Nullwert

The type of the alternating current shown in the figure is known as a *sine wave*. The variations in many a.c. *wavetrains* are not so smooth, and these *complex waves* can be shown to be the sum of two or more sine waves of frequencies that are exact *integral multiples* (*harmonics*) of the *fundamental frequency*.

The three general types of electric circuits are the *series circuit*, the *parallel circuit* and the *series-parallel circuit*. All circuits must contain a source of e.m.f. to establish the difference of potential that makes possible the *current flow*. All *paths* of the circuit lead in *closed loops* from the *high-potential end* of the e.m.f. source to its *zero-potential end*.

alternating current; a.c.	Wechselstrom *m* DIN
closed loop	geschlossene Schleife *f*
complex wave	(hier:) zusammengesetzte Welle *f*
current flow	Stromfluß *m*; Strom *m*
cycle	Periode *f*; Zyklus *m*
direct current; d.c.	Gleichstrom *m* DIN, VDE
frequency IEC, UIP	Frequenz *f* DIN, IEC, UIP

fundamental frequency	Grundfrequenz f
harmonics*	(harmonische) Oberschwingungen $f\,pl$
high-potential end	(hier:) Klemme f mit hohem Potential
integral multiple	ganzzahliges Vielfache n
parallel circuit	Parallelschaltung f; Nebeneinanderschaltung f
path	Weg m
period IEC, UIP	Periodendauer f DIN, IEC, UIP
pure sinusoidal	rein sinusförmig
reversal	Umkehrung f
series circuit	Reihenschaltung f/D/; Serienschaltung f /CH/
series-parallel circuit	Reihenparallelschaltung f
sine wave	(wörtlich:) Sinuswelle f; (besser:) Sinusschwingung f DIN
surge	Spannungsstoß m
temporary	zeitweilig
transient	Übergangserscheinung f
waveform	Wellenform f
wavetrain	Wellenzug m
zero-potential end	(hier:) Klemme f mit Nullpotential

A selection of related terms — **Ausgewählte verwandte Benennungen**

a-c component	Wechselanteil m (einer Mischgröße)
alternation	Halbperiodendauer f
angular frequency UIP	Kreisfrequenz f DIN, UIP
breakdown CEE; puncture	Durschschlag m VDE
coupling	Kopplung f
coupling coefficient UIP	Kopplungsgrad m; Kopplungskoeffizient m UIP
crest factor	Scheitelfaktor m DIN
discharge	Entladung f
brush-~	Büschel~
corona-~	Korona~ DIN
glow-~	Glimm~ DIN
spark-~	Funken~
eddy currents	Wirbelströme $m\,pl$
flashover CEE	Überschlag m VDE
harmonic content	Anteil m an Oberschwingungen; Oberschwingungsgehalt m DIN; Klirrfaktor m DIN
in opposition	in Gegenphase
phase displacement; phase shift	Phasenverschiebungswinkel m; Phasenverschiebung f
phase lag; lag	Nacheilung f
phase lead; lead	Voreilung f
power UIP	Leistung f DIN, UIP
active ~	Wirk~ DIN
apparent ~	Schein~ DIN
reactive ~	Blind~ DIN
power factor	Leistungsfaktor m DIN; Wirkfaktor m DIN

ripple*	Welligkeit f
short circuit	Kurzschluß m VDE
skin effect	Skineffekt m; Stromverdrängung f
unidirectional current	Strom m gleichbleibender Richtung
value	Wert m
peak* ~ ; crest* ~	Größt~ DIN; Maximal~ DIN; Scheitel~ DIN; Gipfel~ IEC; Höchst~ IEC
r.m.s. ~	Effektiv~ DIN, VDE

1.4. Resistance, Inductance and Capacitance
(Widerstand, Induktivität und Kapazität)

The amount of current that will flow in a circuit when a given e.m.f. is applied will be found to vary with what is called the resistance (R) of the material. The *resistivity* is the resistance of a *cube* of the material measuring one centimeter on each edge. The reciprocal of resistance (that is, 1/R) is called *conductance*, and is represented by the symbol G. The *voltage* appearing across a *resistor* in an electric circuit is known as the *voltage drop*.

The *capacitance* may be defined as the *ratio* of the absolute value of the electric charge on either of a pair of *conductors* to the potential difference between them, or, in other words, as the property of a *capacitor* (or a circuit) *to store* electric energy. The ratio of the capacitance with materials other than air between *plates of a capacitor*, to the capacitance of the same capacitor with *air insulation*, is called *relative permittivity* of that particular *insulating material*.

The property of an electric circuit to oppose any change in the current flowing through it is called *inductance*, and *components* of the circuit that produce it are called *inductors*. Specific types of inductance are the *self-inductance* and the *mutual inductance*.

In alternating current circuits the additional factors of reactance (*capacitive reactance* and *inductive reactance*) as well as *impedance* are applied.

air insulation	Luftisolation f
capacitance IEC, UIP	Kapazität f DIN, IEC, UIP
capacitive reactance	kapazitiver Blindwiderstand m
capacitor CEE, IEC	Kondensator m DIN, VDE
component	Bestandteil m
conductance IEC, UIP	Leitwert m DIN, UIP
conductor CEE, IEC	Leiter m DIN, VDE
cube	Würfel m
impedance IEC, UIP	Scheinwiderstand m DIN, UIP
inductance IEC	Induktivität f DIN, VDE
inductive reactance	induktiver Blindwiderstand m
inductor IEC	Spule f
insulating material	Isolierstoff m VDE
mutual inductance IEC, UIP	Gegeninduktivität f UIP
plates of a capacitor	Kondensatorplatten f pl
ratio	Verhältnis n
reactance	Blindwiderstand m DIN
relative permittivity UIP	Dielektrizitätszahl $f(\varepsilon_r)$ DIN, UIP
resistivity* IEC, UIP	spezifischer Widerstand m DIN, UIP
resistor* IEC	Widerstand m (als Bauteil) DIN, VDE
self-inductance IEC, UIP	Selbstinduktivität f DIN; Eigeninduktivität f UIP

to store	speichern
voltage IEC	Spannung *f* DIN
voltage drop IEC	Spannungsabfall *m*
A selection of related terms	**Ausgewählte verwandte Benennungen**
admittance UIP	Scheinleitwert *m* UIP
conductivity* UIP; specific conductance	elektrische Leitfähigkeit *f* DIN, UIP
dielectric constant (obsolete)	Dielektrizitätskonstante *f* DIN, UIP
dielectric loss	dielektrische Verluste *m pl*
dielectric loss angle	Verlustwinkel *m* DIN
dissipation factor (tan δ)	Verlustfaktor (d = tan δ) *m* DIN
electromotive force; e.m.f.	elektromotorische Kraft (EMK) DIN
back ~ ~; counter ~ ~	gegenelektromotorische ~
induced ~ ~	induzierte ~
in opposition	in Gegenphase
in phase	gleichphasig
in quadrature	um 90° phasenverschoben
out of phase	phasenverschoben
permittivity UIP; absolute permittivity IEC; (absolute) electric inductive capacity ISO; electric inductive capacity of the medium ISO; capacitivity	absolute Dielektrizitätskonstante *f* DIN; Dielektrizitätskonstante *f* DIN, UIP; Permittivität *f* pDIN
phase	Phase *f*
resistivity UIP; specific resistance	spezifischer elektrischer Widerstand *m* DIN, UIP
susceptance IEC, UIP	Blindleitwert *m* DIN, UIP
time constant IEC	Zeitkonstante *f* DIN, UIP

Fachsprachliche Anmerkungen

crest value – peak value – amplitude

Die Benennungen *crest value* IEEE und *peak value* IEC sind Synonyme und bezeichnen den größten Betrag einer Mischgröße (*pulsating quantity* IEEE – siehe Bemerkung am Ende des Abschnittes).

Die entsprechenden deutschen Benennungen sind „Größtwert" DIN, „Gipfelwert" IEC, „Höchstwert" IEC oder „Spitzenwert" DIN, wobei die letztgenannte Benennung nur dann zutrifft, wenn der Größtwert der Mischgröße während einer sehr kurzen Zeitspanne im Vergleich zur Periodendauer durchläuft (d. h. vor allem bei impulsartigen Funktionen).

Die Benennung *amplitude* (*of a symmetrical alternating quantity*) IEC, „Scheitelwert" DIN trifft nur für reine Wechselgrößen zu und wird als die „Hälfte der Schwingungsbreite" (*total amplitude*) definiert.

Statt *total amplitude* wird im englischen Schrifttum geläufiger *double amplitude* („Schwingungsbreite" DIN) oder *peak-to-peak value* („Spitze-Spitze") bei impulsartigem Verlauf gebraucht. Im Deutschen wird von „Amplitude" DIN an Stelle „Scheitelwert" meist nur bei sinusförmigem zeitlichen Verlauf gesprochen.

Eine Mischgröße (*pulsating quantity*) liegt vor, wenn einem Gleichanteil (*continous component* IEEE öfter: *d. c. component*) ein Wechselanteil (*alternating component* IEEE öfter: *a. c. component*) überlagert ist.

electric – electrical

Nach der Erläuterung der IEEE bezieht sich das Adjektiv *electric* auf Objekte, die „Elektrizität enthalten, erzeugen, führen, oder führen können, aus Elektrizität be-stehen oder entstehen oder durch Elektrizität betrieben werden". Beispiele: *electric field, electric generator, electric cable, electric brake;* auch *electric eel* („elektrischer Aal", „Zitteraal").

Dagegen bezeichnet *electrical* Gegenstände, die mit Elektrizität verknüpft sind, aber nicht deren Eigenschaften oder Kennzeichen haben. Beispiele: *electrical engineer, electrical insulator, electrical unit.* Die Unterscheidung wird im Fachschrifttum zwar nicht streng, aber ziemlich konsequent eingehalten.

Bei gewissenhafter Sichtung einer großen Anzahl von Wortgruppen mit *electric* und *electrical* (elektrisch) kommt man zu folgenden Regeln:
1. *electric* wird in der überwiegenden Mehrzahl aller Fälle benutzt,
2. *electrical* jedoch in folgenden drei Fällen:
   2.1. bei Benennungen von Allgemeinbegriffen, z. B. *electrical machinery* elektrische Maschinen, *electrical symbols* elektrische Zeichen;
   2.2. vor Substantiven, die mit einem Vokal beginnen, z. B. *electrical energy* (el. Energie), *electrical engineer* Elektroingenieur;
   2.3. wenn die elektrische Eigenschaft des folgenden Substantivs als solche gekennzeichnet werden soll, z. B. *electrical oscillations* el. Schwingungen, *electrical power* el. Leistung.

Ausnahmen bestätigen jedoch die Regel: *electric furnace* elektrischer Ofen.

Dem Adjektiv *electrotechnical* begegnet man im Englischen eigentlich nur im Namen der *IEC (International Electrotechnical Commission);* sonst wird „elektrotechnisch" mit *electrical engineering* umschrieben.

Das Paar *acoustic – acoustical* wurde nach denselben Grundsätzen wie *electric(al)* gebildet, der Unterschied hat sich aber durch Gebrauch verwaschen (dasselbe gilt z. B. auch für *alphabetic – alphabetical.* Für „optisch" und „mechanisch" sind nur *optical* und *mechanical* übriggeblieben. (Ausnahme: *optic nerve* „Sehnerv".)

Das Adjektiv *electronic* existiert nur in der mit *-ic* endenden Form, auch für „nichtelektronische" Objekte, z. B. *electronic engineer.* Mit *-ical* kommt nur die adverbiale Form *electronically* vor.

Für *logic – logical* siehe die fachsprachliche Anmerkung im Kap. 16.

harmonics – harmonic components – harmonic content

In der Praxis wird zwischen *harmonics* IEC („harmonische Oberschwingungen" IEC) und *harmonic components* IEC („harmonische Teilschwingungen", „Harmonische") kein Unterschied gemacht. Laut Definition (IEC) sind aber *harmonics* „Sinusgrößen, deren Frequenz ein ganzzahliges Vielfaches der als Grundfrequenz ausgewählten Größe ist", während *harmonic components* die diesen Oberschwingungen entsprechenden Glieder der Fourier-Reihe bezeichnet.

harmonic content „(Anteil an) Oberschwingungen" ist die Funktion, die durch Beseitigung der Grundfrequenz aus einem nichtsinusförmigen Vorgang entsteht. Das Verhältnis der Effektivwerte der Oberschwingungen und der der nichtsinusförmigen Funktion heißt Oberschwingungsgehalt DIN, zuweilen Klirrfaktor DIN *(distortion factor, relative harmonic content).*

resistance – resistor

Ein lehrreiches Beispiel sinnvoller Terminologie-Normung weist die englische Sprache auf, indem man zur Kennzeichnung einer physikalischen Größe das Ableitungselement *-ance* und zur Kennzeichnung des Trägers einer physikalischen Größe das Ableitungselement *-or* folgerichtig benutzt. Da *capacity* vieldeutig ist, hat man es gewagt, die „Kapazität eines Kondensators" in Anlehnung an *resistance* mit *capacitance* und den Kondensator mit *capacitor* zu benennen. Diese strenge Unterscheidung von

Größe und Träger der Größe wurde mit solcher Folgerichtigkeit durchgeführt, wie sie in keiner anderen der bekannten Sprachen zu finden ist:

Größe	**Träger der Größe**
resistance BS, CEE, IEC	*resistor* BS, IEC
Widerstand DIN, IEC	Widerstand DIN, IEC
	Widerstandsgerät VDE
conductance BS, IEC	*conductor* BS, IEC, CEE
Leitwert DIN, IEC	Leiter DIN, IEC, VDE
capacitance BS, IEC	*capacitor* BS, CEE, IEC
Kapazität DIN, IEC, VDE	Kondensator DIN, IEC, VDE
(zu vermeiden: *capacity*)	(zu vermeiden: *condenser*)
inductance BS, IEC	*inductor* BS, IEC
Induktivität DIN, IEC, VDE	Drossel VDE; Spule
(zu vermeiden: *coefficient of self-induction*)	
reactance BS, IEC	*reactor* BS, IEC
Blindwiderstand DIN	Blindwiderstand DIN

Im Deutschen haben sich Wörter mit der Endung -anz eingebürgert, die jedoch in der Bedeutung vom englischen Muster abweichen: Induktanz für „induktiver Blindwiderstand" *(inductive reactance)*.

resistivity – conductivity

Die Terminologie-Normung hat sich mit dem Ableitungselement *-ivity* befaßt, ohne daß diese Endung an dieselben Wurzeln wie bei *-ance* und *-or* angefügt werden kann; siehe *resistance* und *resistor*.

Eingeführt und durchgesetzt haben sich:
resistivity IEC, UIP spezifischer Widerstand DIN, UIP
conductivity IEC, UIP Leitfähigkeit DIN, UIP

ripple (quantity)

Unter *ripple quantity* IEEE oder kurz *ripple* versteht man allgemein und ohne zahlenmäßige Angabe den Wechselanteil einer Mischgröße, wenn dieser klein gegenüber dem Gleichanteil ist (IEEE, BS). Nach anderen Definitionen (auch IEEE) wird unter *ripple* die ganze Mischgröße (samt Gleichanteil) verstanden.

ripple ratio ist das Verhältnis der Schwingungsbreite einer Mischgröße zu ihrem Mittelwert.

per cent ripple IEEE ist das in % ausgedrückte Verhältnis des Effektivwertes *(r.m.s. value)* des Wechselanteils zum Mittelwert der Mischgröße.

Die deutschen Definitionen entsprechen nicht genau den amerikanischen bzw. britischen, denn:

$$\text{„(effektive) Welligkeit" DIN} = \frac{\text{Effektivwert des Wechselanteils}}{\text{Gleichwert der Mischgröße}}$$

$$\text{„Riffelfaktor" DIN, „Scheitelwelligkeit" DIN} = \frac{\text{Scheitelwert des Wechselanteils}}{\text{Gleichwert der Mischgröße}}$$

Unter Welligkeit wird in der Nachrichtentechnik auch das Verhältnis der Maximal- und der Minimalspannung längs einer Leitung verstanden. Demgemäß muß „Welligkeit" mit *voltage standing-wave ratio* (*VSWR*) übersetzt werden.

2. Electrical and Related Quantities and Units
(Elektrische und verwandte Größen und Einheiten)

2.1. Units and their Dimensions
(Einheiten und ihre Dimensionen)

Every physical measurement *involves* the comparison of two *quantities* of the same nature, e. g. two electrical currents. The same result will be achieved by two different *observers* only if they have agreed to use the same *standard* as one of these quantities. To provide such standards, certain *fundamental units* have been established by *custom*, by *national legislation* and by *international agreement*. The *legal definitions* and/or accepted custom in Great Britain and in the United States agree in most cases, with a few exceptions (e.g. that of the *inch*).

The specification of a *physical quantity* must tell both what standards were used in the measurement and how the quantity *compares* with these standards. The quantity is therefore expressed as the product of a *pure number*, giving the latter *piece of information*, and a unit which gives the former. All quantities which can be expressed in the same units are said to have the same *physical dimensions*. The great majority of physical quantities can be measured in terms of *derived units*, which are defined in terms of the fundamental units.

Which physical quantities should be chosen as fundamental, is a matter of choice and convenience. This is true because most *physical laws* express proportionalities rather than *equalities*. Each time that an arbitrary choice is made, a new *system of units* is established.

to compare	(hier:) sich verhalten
custom	Gewohnheit *f*
derived unit	abgeleitete Einheit *f* DIN
equality	Gleichheit *f*
fundamental unit	Basiseinheit *f* DIN, UIP; Grundeinheit *f*
inch	Zoll *m*
international agreement	internationales Abkommen *n*
to involve	einbeziehen
legal definition	gesetzliche Definition *f*
national legislation	Gesetzgebung *f*
observer	Beobachter *m*
physical dimension	physikalische Dimension *f*
physical law	physikalisches Gesetz *n*
physical quantity UIP	physikalische Größe *f* UIP
piece of information	(hier:) Auskunft *f*
pure number	reine (dimensionslose) Zahl *f*
quantity	Größe *f*
standard	(hier:) Eichnormal *n*
system of units	Einheitensystem *n* DIN

2.2. Systems of Units
(Einheiten-Systeme)

During the early development of many parts of science, all that was possible or required was to measure the *properties* of various *substances*. As the *art of measurement* progresses, such terms are usually *redefined* in such a way as to acquire *absolute meanings* in a system of units.

A system of units should satisfy the following three basic requirements: (1) the *size of units* should be such that the numbers encountered in practical problems are reasonable, (2) the relative size of the units should be such that they *fit together* without *annoying conversion factors*, and (3) the *multiples* and *submultiples* of each unit should be on a decimal basis, to facilitate *computation*.

The traditional English system fails to satisfy requirements (2) and (3). Similarly, the older centimetre-gramme-second, or cgs-system, is quite deficient in (2) and somewhat in (1). At present the generally accepted system is the *International System of Units* based on six *basic units* (m, kg, s, A, K, cd) which includes the Giorgi or MKSA system. These units are called *SI-Units* (the designation "SI" was adopted in 1960 by the eleventh "Conférence Générale des Poids et Mesures").

A detailed survey of quantities and units used in *electrical technology* may be found in the annexed table. The table also includes the respective *letter symbols* for quantities and units.

Symbols used for denoting quantities are single letters of the Latin or Greek alphabet, sometimes with *subscripts* or other *modifying signs*. For indicating the *vector character* of a quantity, *bold-face italic type* is recommended (for example ***H***). If such a type is not available, an *arrow* may be placed over the letter symbol (for example $\vec{H}$).

Symbols for units are written in *lower-case letters* (meter, gram, etc.), except the first letter when the name of the unit is derived from a *proper name* (Volt, Ampere, Ohm, etc.). They remain unaltered in the plural and are written without a final stop (period).

absolute meaning	absolute Bedeutung *f*
annoying conversion factor	umständlicher Umrechnungsfaktor *m*
arrow	Pfeil *m*
art of measurement	Meßtechnik *f*
basic unit	Basiseinheit *f* DIN, UIP; Grundeinheit *f*
bold-face italic type	fettgedruckte Kursivschrift *f*
computation	Rechenverfahren *n*
electrical technology	(hier:) Elektrotechnik *f*
to fit together	zusammenpassen
International System of Units	Internationales Einheitensystem *n* DIN
letter symbol	Buchstabenzeichen *n*
lower case letter	kleiner Buchstabe *m*
modifying sign	(hier:) Bestimmungszeichen *n*
multiple	Vielfache *n*
proper name	Eigenname *m*
property	Eigenschaft *f*
to redefine	neu definieren
SI-unit	SI-Einheit *f* DIN
size of units	Größe *f* der Einheiten
submultiple	(hier:) Bruch *m*; dekadischer Teil *m*
subscript	(unterer) Index *m*
substance	Stoff *m*
vector character	Vektorcharakter *m*

A selection of related quantities and units **Ausgewählte verwandte Größen und Einheiten**

Geometry and Kinematics **Geometrie und Kinematik**

area; surface area	Flächeninhalt *m*
volume	Rauminhalt *m*
time	Zeit *f*

time of one cycle; period	Periodendauer f
time constant	Zeitkonstante f
rotational frequency*	Drehzahl f
angular frequency	Kreisfrequenz f
angular velocity	Winkelgeschwindigkeit f
angular acceleration	Winkelbeschleunigung f
(linear) speed; velocity	Geschwindigkeit f
(linear) acceleration	Beschleunigung f

Dynamics / Dynamik

mass	Masse f
density; mass density	Massendichte f
momentum*	Wucht f
(dynamic) moment of inertia	Trägheitsmoment n
force*	Kraft f
weight	Gewicht n
weight density; specific weight	Wichte f
torque	Drehmoment n
pressure	Druck m
work	Arbeit f
energy	Energie f
power	Leistung f
efficiency	Wirkungsgrad m

Thermodynamics / Thermodynamik

thermodynamic temperature; absolute temperature (degree kelvin*)	absolute Temperatur f
heat; quantity of heat	Wärmemenge f
heat capacity	Wärmekapazität f
specific heat capacity	spezifische Wärme(kapazität) f

Radiation / Strahlung

radiant energy	Strahlungsenergie f
radiant flux; radiant power	Strahlungsfluß m
radiant intensity	Strahlungsstärke f
radiance	Strahldichte f
irradiance	Bestrahlungsstärke f

Light / Licht

luminous intensity	Lichtstärke f
luminous flux	Lichtstrom m
quantity of light	Lichtmenge f
luminance	Leuchtdichte f
illuminance; illumination	Beleuchtungsstärke f

Electrical and Magnetic Quantities, Units and Constants
(Elektrische und magnetische Größen, Einheiten und Konstanten)

Quantity (Größe)				Corresponding Unit (Entsprechende Einheit)				
IEC			DIN	SI-unit (Si-Einheit)			Other Unit or Designation (Andere Einheit oder Bezeichnung)	
Term (Benennung)	Symbol		Term (Benennung)	Benennung	Zeichen		Benennung	Zeichen
	Chief	Reserve						
(electric) charge; quantity of electricity	Q		elektrische Ladung; Elektrizitätsmenge	coulomb	C		ampere hour	Ah
surface density of charge	σ		Oberflächenladungsdichte	coulomb per square metre	C/m^2			
volume density of charge	ϱ	η	Raumladungsdichte	coulomb per cubic metre	C/m^3			
electric field strength	E	K	elektrische Feldstärke	volt per metre	V/m			
(electric) potential	V	φ, Φ	elektrostatisches Potential	volt	V			
potential difference; tension; voltage	U	V	elektrische Spannung	volt	V			
electromotive force	E		elektromotorische Kraft	volt	V			
electric flux	ψ		elektrischer Fluß	coulomb	C			
electric flux density; electric displacement	D		elektr. Verschiebung; elektrische Flußdichte	coulomb per square metre	C/m^2			

Quantity (Größe)				Corresponding Unit (Entsprechende Einheit)			
IEC			DIN	SI-unit (Si-Einheit)		Other Unit or Designation (Andere Einheit oder Bezeichnung)	
Term (Benennung)	Symbol		Term (Benennung)	Benennung	Zeichen	Benennung	Zeichen
	Chief	Reserve					
capacitance	C		elektrische Kapazität	farad	F		
permittivity; absolute permittivity; (capacitivity)	ε	ϵ	(absolute) Dielektrizitätskonstante	farad per metre	F/m		
relative permittivity	ε_r	ϵ_r	Dielektrizitätszahl	(dimensionless)			
electrization $[E_i = (D/\varepsilon_o) - E]$	E_i	K_i	Elektrisierung	volt per metre	V/m		
electric polarization $[P = D - \varepsilon_o E]$	P	D_i	elektrische Polarisation	coulomb per square metre	C/m^2		
electric dipole moment	p	p_e	elektrisches Moment	coulomb metre	C.m		
electric current	I		elektrische Stromstärke	ampere	A		
current density	J	S	elektrische Stromdichte	ampere per square metre	A/m^2		
linear current density	A	α	Oberflächenstromdichte	ampere per metre	A/m		
magnetic field strength	H		magnetische Feldstärke	ampere per metre	A/m	oersted	Oe; N/Wb

magnetic potential difference	U, U_m	$\mathfrak{U}$	magnetische Spannung; V	ampere		A	J/Wb
magnetomotive force $[F = \oint H_s \, ds]$	F, F_m	$\mathfrak{F}$	magnetomotorische Kraft	ampere	ampere-turn; gilbert	A	At Gb
magnetic flux density; (magnetic induction)	B		magnetische Induktion; magnetische Flußdichte	tesla	gauss	T	Gs; Wb/m^2
magnetic flux	Φ		magnetischer Induktionsfluß; magnetischer Fluß	weber	maxwell	Wb	Mx; V$_s$
magnetic vector potential	A		(magnetisches) Vektorpotential	weber per metre		Wb/m	
self inductance	L		Selbstinduktionskoeffizient	henry		H	
mutual inductance	M, L_{mn}		Gegeninduktionskoeffizient	henry		H	
coupling coefficient [for example $k = = L_{12}(L_1 L_2)^{-\frac{1}{2}}$]	k	$\varkappa, \kappa$	Kopplungsgrad	(dimensionless)			
leakage coefficient $[\sigma = 1 - k^2]$	σ		Streukoeffizient	(dimensionless)			
permeability; absolute permeability	μ		absolute Permeabilität	henry per metre		H/m	
relative permeability	μ_r		Permeabilitätszahl	(dimensionless)			
magnetic susceptibility	$\varkappa, \kappa$		magnetische Suszeptibilität	(dimensionless)			

Quantity (Größe)				Corresponding Unit (Entsprechende Einheit)				
IEC			DIN	SI-unit (Si-Einheit)			Other Unit or Designation (Andere Einheit oder Bezeichnung)	
Term (Benennung)	Symbol		Term (Benennung)	Benennung	Zeichen		Benennung	Zeichen
	Chief	Reserve						
magnetic moment*; (magnetic area moment)	m		(Ampèresches) magnetisches Moment	ampere metre squared	$A.m^2$			
magnetization $[H_i = (B/\mu_o) - H]$	H_i, M		Magnetisierung	ampere per metre	A/m			
intrinsic magnetic flux density; magnetic polarization; intrinsic induction $[B_i = B - \mu_o H]$	B_i, J		magnetische Polarisation	tesla	T		Gauss	G
magnetic dipole moment*	j		(Coulombsches) magnetisches Moment	newton metre squared per ampere	$N.m^2/A$		weber metre	Wb.m
resistance	R		elektrischer Widerstand	ohm	Ω			
resistivity	ϱ		spezifischer elektrischer Widerstand	ohm metre	$\Omega.m$			
conductance	G		elektrischer Leitwert	siemens	S		mho	mho
conductivity $[\gamma = 1/\varrho]$	γ, σ		elektrische Leitfähigkeit	siemens per metre	S/m			
reluctance	R, R_m	$\mathfrak{R}$	magnetischer Widerstand	reciprocal henry	H^{-1}			

permeance $[\Lambda = 1/R_m]$	Λ		magnetischer Leitwert	henry	H
impedance	Z		Scheinwiderstand	ohm	Ω
reactance	X		Blindwiderstand	ohm	Ω
quality factor	Q		Gütefaktor	(dimensionless)	
loss angle	δ		Verlustwinkel	radian	rad
admittance $[Y = 1/Z]$	Y		Scheinleitwert	siemens	S
susceptance	B		Blindleitwert	siemens	S
active power	P		Leistung	watt	W
reactive power	Q	P_q	Blindleistung	var	var
apparent power	S	P_s	Scheinleistung	voltampere	VA
Poynting vector	S		Poyntingscher Vektor	watt per square metre	W/m²
phase difference; phase displacement	φ, Φ	ϑ, θ	Phasenverschiebung	(dimensionless)	
number of turns (in a winding)	N		Windungszahl	(dimensionless)	
number of phases	m		Phasenzahl	(dimensionless)	
number of pairs of poles	p		Polpaarzahl	(dimensionless)	

Constants (Konstanten)

IEC		DIN	
Term (Benennung)	Zeichen	Benennung	Einheit
elementary charge	e	Elementarladung	C
electric constant; permittivity of vacuum	$\varepsilon_o, \in_o$	elektrische Feldkonstante; (elektrische) Verschiebungskonstante; (elektrische) Influenzkonstante	F/m
magnetic constant; permeability of vacuum	μ_o	magnetische Feldkonstante; (magnetische) Induktionskonstante; Leerinduktion	H/m
velocity (speed) of propagation of electromagnetic waves [in vacuo c_o]	c	Ausbreitungsgeschwindigkeit von elektromagnetischen Wellen (im Vakuum)	m/s
characteristic wave impedance	Γ_o	Wellenwiderstand des leeren Raumes; Feldwellenwiderstand des freien Raumes	

2.3. UK Goes Metric
(Übergang zum metrischen System im Vereinigten Königreich)

According to the Institution of Electrical Engineers, *metric values* are to be reckoned as equivalent to the established *Imperial system*.

In 1969, a detailed programme for the *metrication* of the British electrical industry was published. A *sector-by-sector timetable* shows the dates for various stages of metrication in the main groups of the industry. It is admitted that some manufacturers will *switch over* in advance of the *scheduled target dates*. But some, especially those concerned with the North American market may want to *delay going metric*.

Regulations are also affected, but full metrication of the regulations is not to be considered until a new edition is prepared. Nevertheless several exceptions from strict SI units have been approved. Thus, the kilowatt hour is not to be *superseded* by the megajoule. Although in the SI system the second is the basic unit of time, the British Standards Institution has included the kilowatt hour in its latest publication on SI units as a *permissible alternative* to the joule because of its *widespread use* in the electrical industry.

Another exception is accepted for pressure. The Newton per m² is the *preferred* SI unit, but for *pressure vessels* the bar and its multiples will be used (1 bar = $10^5 N/m^2$). The hectobar is to be used for indicating the *strength* of metallic materials.

(From: Electronics Weekly, 1969)

to delay going metric	den Übergang zum metrischen System verzögern
Imperial system	Empire-Einheitensystem *n*; gesetzliches Einheitensystem *n* (in GB)
metrication; metrification	Übergang *m* zum metrischen System
metric values	(hier:) metrische Einheiten *fpl*

permissible alternative	zulässige zweite Möglichkeit *f*
preferred	Vorzugs-
pressure vessel	Druckgefäß *n*
regulations	Vorschriften *fpl*
scheduled target dates	geplante Termine *mpl*
sector-by-sector timetable	nach Industriesektoren aufgestellter Zeitplan *m*
strength	Festigkeit *f*
to supersede	ersetzen
to switch over	(hier:) sich umstellen
timetable	Stundeneinteilung *f*; (hier:) Zeitplan *m*
widespread use	verbreiteter Gebrauch *m*

Fachsprachliche Anmerkungen

degree kelvin – Celsius degree

Die Si-Einheit für *thermodynamic temperature* UIP („Temperatur" DIN, UIP) ist seit 1967 das Kelvin DIN.

Bei Angabe von Celsius-Temperaturen DIN, UIP (*temperature* UIP) wird die Einheit Grad Celsius (Einheitszeichen °C DIN) (*Celsius degree, symbol* °C IEC) verwendet, die gleich der Einheit Kelvin (K) ist.

Die Einheit der Temperaturdifferenzen und -intervalle ist ebenfalls das Kelvin (K) [*kelvin* (K)]; daneben gebräuchlich ist Grad DIN (*degree* IEC), das als grd DIN, im ausländischen Schrifttum als *deg* (ohne Punkt) IEC abgekürzt wird. (Nicht zu verwechseln mit *grade* (Kurzzeichen *gr*) = englische und französische Bezeichnung für die Winkeleinheit „Gon".)

Obwohl die CGPM (Conférence générale des poids et mesures) schon vor etwa 25 Jahren das Wort *centigrade* zugunsten von *degree Celsius* verstoßen hat, findet man in der amerikanischen Literatur immer noch öfter die erste Bezeichnung; am häufigsten stößt man allerdings auf die uralte Temperatureinheit *deg F* (*degree Fahrenheit*).

kilo – mega – pico – micro

Obwohl sich die internationale Normung der Vorsatzzeichen für Einheiten allmählich durchsetzt, findet man noch sehr oft (insbesondere in amerikanischen Zeitschriften für Techniker) Vorsatzzeichen, die den Normen nicht entsprechen:
1. für *kilo* wird das Zeichen *K* (statt *k*) benutzt;
2. für *mega* das Zeichen *m* (statt *M*);
3. für *pico* das Zeichen µµ, *uu*, *UU*, *MM* (= „*micromicro*") statt *p*;
4. bisweilen auch µ statt µ*m* (micrometer, 10^{-6}m); diese Schreibweise war übrigens auch in vielen europäischen Ländern üblich;
5. kombinierte Vorsatzzeichen, z. B. „*kmc*" („*kilomegacycle*") statt *GHz* (*gigahertz*).
6. auch in handschriftlichen Ausarbeitungen findet man *m* oder *M* (statt µ) z. B. in µV/m.

kilogramme-force

Für den deutschen Einheitennamen „Kilopond" DIN wird im angelsächsischen Fachschrifttum nur der Name *kilogramme-force* IEC bzw. *kilogram-force* ANSI, IEEE mit dem Einheitszeichen *kgf* verwendet; 1 *kgf* gleicht also 1 kp oder ungefähr 9,8 Newton.

Dementsprechend benutzt man das Zeichen *f* auch in z. B. *lbf* (*pound-force*), *lbf/in²* (*pound-force per square inch*) und *tonf* (= *ton-force*). Für die Krafteinheit im britischen Einheitensystem (*foot, lbf, s*) wurde manchmal die Benennung „*poundal*" benutzt, wobei 1 *poundal* = 1 *lb*. *foot*/s^2 = 0,014 kp.

Im Deutschen hört man zuweilen die Aussprache [kilopaund] für Kilopond, wohl aus dem Irrglauben heraus, es stamme aus dem Englischen. Das Wort „Pond" ist 1934 von Fritz Hoffmann (PTR = Physik.-Techn. Reichsanstalt) in Anlehnung an *pondus* (= Gewicht) geprägt worden. Im Englischen konnten sich pond und kilopond nicht durchsetzen, weil sie zu leicht mit *pound* verwechselt werden können.

mho – μmho

Die Röhrensteilheit *mutual conductance* BS, *transconductance* /USA/, die in Europa (außer Großbritannien) fast ausschließlich in A/V oder S angegeben wird, wird in den angelsächsischen Ländern in *mho* gemessen; *mho* ist der reziproke Wert eines Ohms ($1/\Omega$). Statt μ*mho* (10^{-6} *mho*) findet man oft *umho* (oder sogar *UMHO*).

Das Wort *mho* als Benennung der Einheit des Leitwertes ist von Lord Kelvin geprägt worden, indem er das Wort „ohm" umkehrte. Da in Deutschland „mho" als eine Verstümmelung des Namens Ohm empfunden und deshalb abgelehnt wurde, hat 1908 der Ausschuß für Einheiten und Formelgrößen (AEF) die Benennung „Siemens" und das Zeichen S vorgeschlagen.

moment – momentum

Diese beiden Begriffe müssen streng auseinandergehalten werden: *moment* ist „Moment" DIN (mechanisches, magnetisches usw.); *momentum* ist dagegen „Impuls" IEC, „Bewegungsgröße". Die Wortverbindung *moment of momentum* bezeichnet den Drehimpuls. Die Benennung *moment of a couple* ISO entspricht der landläufigen Benennung *torque* BS, IEC.

speed of rotation – rotational frequency – rotational speed

Die in der angelsächsischen Fachliteratur am häufigsten vorkommenden Benennungen für „Drehzahl" sind *revolutions per second* (*r.p.s.*) oder *revolutions per minute* (*r.p.m.*), *number of revolutions* und *speed of rotation*. Da es sich hier jedoch im Grunde um eine Frequenz (mit der Dimension z. B. s^{-1}) handelt, ist eine genauere und empfohlene Benennung *rotational frequency* IEC.

Rotational frequency of electrons sollte aber mit „Elektronenumlaufsfrequenz" übersetzt werden.

Die Bezeichnung *rotational speed* ist zweideutig: sie kann sich entweder auf die Drehzahl (z. B. in s^{-1}), oder auf die Umlaufsgeschwindigkeit (z. B. $m.s.^{-1}$) beziehen.

standard terminology of electrical quantities

Die genormte Terminologie der ISO enthält einige Benennungen, die in anderen Normwerken (IEC, BS, ANSI) nicht vorkommen und auch im Fachschrifttum ziemlich selten sind:

ISO	IEC und übliche Fachterminologie	DIN
periodic time	*period*	Periodendauer
moment of a couple	*torque*	Drehmoment
customary temperature	*temperature*	Temperatur
(*electric*) *tension*	*voltage*	(elektrische) Spannung
flux of displacement	*electric flux*	elektrischer Fluß
electromagnetic moment	*magnetic moment*	Ampèresches magnetisches Moment

temperature coefficient

Die Benennung *temperature coefficient* (abgekürzt *T.C.*) „Temperaturkoeffizient", „TK") ist nicht vollständig definiert, solange nicht die sich ändernde Größe angegeben ist.

Die Angabe dieser Größe fehlt aber im englischen Fachschrifttum oft, wenn sie keinem Zweifel unterliegt, sollte aber bei genauer Übersetzung ergänzt werden, z. B.

temperature coefficient (richtig *resistance temperature coefficient*) = „Widerstandstemperaturkoeffizient".

Electrical Materials
(Werkstoffe der Elektrotechnik)

1 Conductors
(Leiter)

An electrical *conductor* is a material which, when placed between *terminals* having a difference of electrical potential, will readily permit the passage of electric current. Different materials have different degrees of *conductivity*. The best conductors are the metals, such as silver, *copper*, platinum, *mercury*, etc., but *non-metallic substances* such as *carbon* or *saline solutions* also are sufficiently *conductive*.

Some types of *high-resistivity* conductors are known as *semiconductors* (they are discussed in detail in Chapter 12 of this book).

Silver ist the best of all conducting materials but high cost prevents its use for general purposes. Copper, in the form of *pure copper* or *cadmium-copper* conductors, is the most generally used of all conducting materials. Aluminium with its conductivity of about 60 per cent and *density* of about 30 per cent of that of copper is used in *overhead lines*, and its *mechanical strength* is usually increased by *stranding* over a stranded *steel core*. Some of the other conducting materials are *electrical resistance alloys*, *contact metals* (e.g. *tungsten* or *noble metals*, such as gold and the *platinum group*), *silicon carbide, fusible metals* and *alloys*, etc.

Conductors are manufactured in various forms and shapes. These may be *wires*, *cables, flat straps*, square or rectangular *bars, angles, channels* or special designs for particular requirements.

A wire is a slender rod or *filament* of *drawn metal* (mainly *annealed copper*, steel, aluminium or phosphor bronze). Wires are sized by the areas in *circular mils* or in *wire gauges* but there is a tendency to abandon *gauge numbers* entirely and specify wire sizes by the diameter in mils.

Electrical conductors are *bare* or *insulated* by *rubber compounds, varnished cambric, asbestos*, paper, silk, cotton, *enamel* or various types of *plastics*. Conductors are sometimes protected by *fibrous braids, lead sheaths* or *metallic armours*.

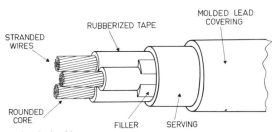

Fig. 3–1. Lead sheathed cable.
(Bleimantelkabel.)

filler	Füllmasse
molded lead covering	gepreßter Bleimantel
rounded core	Rundseele
rubberized tape	gummiimprägniertes Isolierband
serving	Umwicklung
stranded wires	verseilte Drähte

A cable is either a stranded conductor (*single-conductor cable*) or a combination of conductors insulated from one another (*multiple-conductor cable*). The term cable is a general one, a small cable is more often called a *stranded wire* or *cord*. An example of a lead-sheathed cable is shown in Fig. 3–1.

alloy	Legierung *f*
angle	Winkelprofil *n*
annealed copper	geglühtes Kupfer *n*
asbestos	Asbest *m*
bar	Stange *f*
bare	blank
cable	Kabel *n* VDE
cadmium copper	Kupferkadmiumlegierung *f*
carbon	Kohlenstoff *m*
channel	U-Profil *n*
circular mil	Kreisfläche mit 1 mil Durchmesser
conductive	leitend
conductivity IEC, UIP	(elektrische) Leitfähigkeit *f* DIN, UIP
conductor IEC	Leiter *m* DIN, VDE, IEC
contact metal	Kontaktmetall *n*
copper	Kupfer *n*
cord CEE	Schnur *f*; Leitungsschnur *f* VDE
density	Dichte *f* DIN
drawn metal	gezogenes Metall *n*
electrical resistance alloy	Widerstandslegierung *f*
enamel	Emaille *f*; Email *n*
fibrous braid	faserige Umflechtung *f*
filament	Faden *m*
flat strap	flache Stange *f*; flaches Flechtband *n*
fusible metal	leicht schmelzbares Metall *n*
gauge* number	(hier): Drahtlehren-Nummer *f*
high-resistivity	hoher spezifischer (elektrischer) Widerstand *m*
to insulate*	isolieren
lead* sheath	Bleimantel *m* VDE
mechanical strength	Festigkeit *f*
mercury	Quecksilber *n*
metallic armour	Metallbewehrung *f* VDE
mil	10^{-3} Zoll
multiple conductor cable	Mehrleiterkabel *n*
noble metal	Edelmetall *n*
non-metallic substance	Nichtmetall *n*
overhead line	Freileitung *f*
plastic	Kunststoff *m*
platinum group	(hier:) Metalle *npl* der Platingruppe
pure copper	Reinkupfer *n*
rubber compound	gummihaltige Verbindung *f*
saline solution	Salzlösung *f*
semiconductor IEC	Halbleiter *m* IEC
silicon carbide	Siliziumkarbid *n*
single-conductor cable	Einzelleiterkabel *n*
steel core	Stahlseele *f*

to strand	verseilen
stranded wire	Litzendraht *m* VDE
terminal IEC	Klemme *f* VDE
tungsten	Wolfram *n*
varnished cambric	Lackbaumwolle *f*
wire	Draht *n*
wire gauge	Drahtlehre *f*

A selection of related terms — **Ausgewählte verwandte Benennungen**

aluminium /GB/; aluminum /USA/	Aluminium *n*
bundle conductor	Bündelleiter *m*
bus-bar	Sammelschiene *f*
cable laying	Kabellegung *f*
cable lay-up	Kabelverseilung *f*
composite conductor	Doppelmetalleiter *m*
(flexible) cord CEE	Leitungsschnur *f*
hard-drawn	kaltgezogen
laminations	Transformatorblech *n*
lay (of a cable)	Drallänge *f* (eines Kabels)
silver-plated	versilbert
solid wire	Volldraht *m*

2 Dielectrics
(Dielektrische Werkstoffe)

Any *dielectric* is an electric insulator but experience has demonstrated the value of certain *solids* for practical use. Some of the best insulators arranged in accordance with their respective insulating properties are dry air, glass, *built-up mica*, and rubber. Other important members of the insulator family are ceramics and *synthetic resins* (also termed plastics), which can be divided into *thermo-plastic* and *thermo-setting materials*. Oil and air are often used where very high voltages are employed. Many interesting applications exist for the *silicone* which appears in the form of fluids, greases, emulsions, resins and rubbers.

With *solid insulators*, mechanical strength is often a consideration; as are also *flexibility*, high *surface resistance* or the possibility of being *machined* and *moulded*.

Thermal effects often seriously influence the choice of insulating materials, some of the principal features being *melting point* (in contrast to *freezing point*), *softening temperature*, *ageing* due to heat, and their *thermal resistivity* expressed, for example, as *ignitability*, *incombustibility*, ability to *self-extinguish* when *ignited*, or *flash point*.

The primary requirement of an insulator, however, concerns its *insulating strength*, i.e. the maximum voltage per *unit thickness* which the material will sustain without *electric breakdown* or *sparkover*.

Other important properties of insulating materials concern their *permittivity* and *dielectric losses* which may be expressed in terms of *dielectric loss angle* or *dielectric loss factor* (*tan δ*).

ageing	Altern *n*
built-up mica*	Mikanit *n*
dielectric	Dielektrikum *n*
dielectric loss angle	Verlustwinkel *m* DIN
dielectric losses	dielektrische Verluste *mpl*
dielectric loss factor	Verlustfaktor *m* DIN, VDE
electric breakdown	elektrischer Durchschlag *m*
flash point	Flammpunkt *m*

flexibility	Biegsamkeit *f*
freezing point	Gefrierpunkt *m*
ignitability	Entzündbarkeit *f*
to ignite	zünden
incombustibility	Nichtbrennbarkeit *f*
insulating* strength	Isolationsfestigkeit *f*
to machine	maschinell (spanabhebend) bearbeiten
melting point	Schmelzpunkt *m*
mica	Glimmer *m*
to mould /GB/; to mold /USA/	formen, pressen
permittivity IEC	Dielektrizitätskonstante *f* DIN; Permittivität *f* pDIN
to self-extinguish	sich selbst löschen
silicone	Silikon *n*, Silicon *n*
softening temperature	Erweichungstemperatur *f*
solid	fester Körper *m*
solid insulator	fester Isolator *m*
sparkover	Überschlag *m* VDE
surface resistance	(hier:) Verschleißfestigkeit *f*
synthetic resin	Kunstharz *n*
tan δ	tan δ = d DIN
thermal resistivity	Wärmebeständigkeit *f*; Wärmewiderstand *m*
thermo-plastic material	thermoplastischer Kunststoff *m*; Thermoplast *m*
thermo-setting material	Duromer *n*, Duroplast *m*, hitzehärtbarer Kunststoff *m*
unit thickness	Einheitsdicke *f*
A selection of related terms	**Ausgewählte verwandte Benennungen**
disruptive breakdown	Durchschlag *m*
disruptive voltage	Durchschlagsspannung *f*
electric stress	dielektrische Beanspruchung *f*
flashover	Überschlag *m* VDE
moisture absorption	Wasseraufnahmevermögen *n*
resistivity	spezifischer Widerstand *m*
surface ~	Oberflächen~
volume ~ ; bulk ~	Durchgangs~
strength	Festigkeit *f*
bending ~	Biege~ DIN
compressive ~	Druck~ DIN
electric ~	elektrische ~
impact ~	Schlag~
shearing ~	Schub~ DIN
tensile ~	Zug~ DIN; Zerreiß~ DIN

3.3 Magnetic Materials
(Magnetische Werkstoffe)

The *electrical technology* is dependent on four basic types of materials – good conductors and high-resistivity conductors, insulators and materials termed "magnetic". (The term "magnetic", in addition to its general meaning of *"pertaining to* magnetism", carries a technical implication of "ferromagnetic".)

The main ferromagnetic materials are iron, nickel, and cobalt. All three elements, and many others which themselves are non-magnetic, are used in special alloys in which certain *magnetic characteristics* are developed to a high degree.

One of the basic characteristics of ferromagnetic materials is *hysteresis*. The simplest method of illustrating the property of hysteresis is by graphical means such as the *hysteresis loop* shown in Fig. 3–2.

In this figure the *magnetizing force* is indicated along the plus and minus H axis, and the *magnetic flux density* is indicated along the plus and minus B axis. The intensity of the magnetizing force, H, applied by means of a *current-carrying coil*, is varied uniformly through one cycle of operation, starting at zero. The force, H, is increased in the positive direction, and during this time the flux density, B, increases from zero to point A.

Fig. 3–2. Hysteresis loop (magnetization, or B-H curve) of a ferromagnetic material.
(Hystereseschleife eines ferromagnetischen Stoffes.)

B axis – flux density	magnetische Induktion
H axis – magnetizing force	Magnetisierung
OA – rise path	Neukurve (jungfräuliche Kurve)
OB, OE – retentivity	Remanenz
OC, OF – coercive force	Koerzitivfeldstärke

If H is decreased to zero, the decreasing curve of flux density does not return to zero *via* its *rise path*. Instead, it returns to point B. The magnetic flux indicated by the length of line OB represents the *retentivity* of the *magnetic substance*.

The value of the *residual*, or *remaining*, *magnetism* when H has been reduced to zero, depends on the substance used and the degree of flux density attained.

If current is now *sent* through the coil in the opposite direction, so that the intensity of the magnetizing force becomes –H, the force will have to be increased to point C before the residual magnetism is reduced to zero. The magnetizing force, OC, necessary to reduce the residual magnetism to zero is called *coercivity*.

If the magnetizing force is continued the curve descends from C to D, magnetizing the *sample* of magnetic material with the *opposite polarity*. If the magnetizing force is reduced again to zero, the flux density is reduced to point E. The magnetic flux indicated by the length of line OE represents the retentivity of the magnetic substance, as did line OB.

If the current through the coil is again *reversed*, the *magnetization curve* moves to zero when the magnetizing force is increased to point F.

There are two distinct groups of ferromagnetic materials, those which are easily demagnetized, and those which are not. These are often designated as *soft* and *hard magnetic materials*. An excellent soft magnetic material is iron. It attains the highest *saturation value* of magnetic induction with the exception of certain cobalt alloys. However, its maximum *permeability* is surpassed by other materials. When used as a *core* in a field induced by a.c., the so-called *iron losses* are relatively high. These losses consist of energy losses related to the area of the hysteresis loop, and of *eddy-current losses*.

In the *communications field*, where the use of very high frequency currents greatly aggravates the problem of eddy-current losses, various powdered materials are used. They are produced *in powdered form*, given a very thin *insulating film*, and *compacted* under high pressure into a suitable core shape. Some of the other special magnetic materials intended for communications are the *ferrites* and the *mumetal alloys*.

coercivity	Koerzitivfeldstärke *f* DIN; Koerzitivkraft *f*
communications field	(hier:) Nachrichtentechnik *f*
to compact	kompakt machen
core	Kern *m*
current-carrying coil	stromdurchflossene Spule *f*
eddy current losses	Wirbelstromverluste *mpl*
electrical technology*	Elektrotechnik *f*
ferrite	Ferrit *n*
hard magnetic material	hartmagnetischer Stoff *m*
hysteresis	Hysterese *f*
hysteresis loop	Hystereseschleife *f* DIN
in powdered form	pulverförmig, in Pulverform
insulating film	Isolierschicht *f*
iron losses	Eisenverluste *mpl*
magnetic characteristic	(hier:) magnetische Eigenschaft *f*
magnetic flux density	magnetische Induktion *f* DIN
magnetic substance	magnetischer Stoff *m*
magnetization curve	Magnetisierungskurve *f*
magnetizing force	Magnetisierung *f*
mumetal alloy	Mumetallegierung *f*
opposite polarity	entgegengesetzte Polarität *f*
permeability IEC	Permeabilität *f* DIN
to pertain to	mit etw. verbunden sein; gehören zu
remaining magnetism; residual magnetism	verbleibende Magnetisierung *f*
retentivity	Remanenz *f*
to reverse	umkehren
rise path	Anstiegsverlauf *m* (s. jedoch auch Legende zur Abb. 3–2)
sample	Materialprobe *f*
saturation value	Sättigung *f*
to send	(hier:) führen
soft magnetic material	weichmagnetischer Stoff *m*
via	über

Fachsprachliche Anmerkungen

built-up (material)

to build „bauen, errichten" wird meistens in der Wortgruppe "*to build up*" verwendet, wo es dann heißt „aufbauen, gestalten, zusammenstellen und -setzen".

built up heißt somit wörtlich „aufgebaut", „zusammengesetzt". Solche Werkstoffe entstehen z. B. durch Verleimen, Sinterung u. ä. Verfahren aus zerkleinerten Ausgangsmaterialien (Fasern, Flocken usw.). Im gleichen Sinne werden auch die Adjektive *reconstructed* und *recovered* benutzt. So heißt z. B. Mikanit *built-up mica*, *reconstructed mica*, *recovered mica*, oder kurz *remica*, seltener *micanite*. Statt *built-up material* (Verbundwerkstoff) sagt man auch *composite material*.

insulation – isolation

In den Fällen, wo die deutsche Fachsprache die Wörter „isolieren", „Isolator" DIN, VDE benutzt, ist die richtige Übersetzung *to insulate*, bzw. *insulator* IEC. Dies gilt sowohl für die elektrische als auch für die thermische und akustische Isolierung. Außerdem kennt die englische Fachsprache die Wörter *to isolate, isolator* IEC, jedoch im Sinn des Begriffes „trennen", „Trenner" DIN, IEC, „Trennschalter" DIN, IEC, VDE; praktisch kommt es nur in Verbindung *isolator switch* oder *isolation switch* BS „Trennschalter" (Starkstromtechnik) und als *isolator* „Richtungsisolator, Richtleiter" (Mikrowellentechnik) vor. Bisweilen findet man auch *isolated neutral system* IEC („isoliertes Netz, Netz mit isoliertem Sternpunkt") neben den viel häufiger benutzten Benennungen *ungrounded system* ANSI, *insulated (supply) system* BS. *to isolate* heißt „stromlos abschalten, trennen", nicht „isolieren"!

In der Funkstörungs-Meßtechnik wird der 50-Hz-Tiefpaß zwischen dem Starkstromnetz (*mains*) und der Netznachbildung ÖVE, VDE (*artificial mains network* CISPR, IEC) wie folgt genannt: *isolating unit* CISPR, *isolating network* BS, *isolation network* CISPR und *radio-frequency isolating filter* IEC.

lead

Das in der Fachliteratur oft gebrauchte Wort *lead* hat eine Fülle von Bedeutungen, die für den Elektrotechniker von Interesse sein mögen. Abgesehen von *lead* „Blei", lies: [led] sind alle Bedeutungen vom Begriff „führen" abgeleitet; so ist *lead* [li:d]
1. (Strom)zuführung, Leiter, Zuleitung;
2. Voreilung (Phasenvoreilung, *phase lead*);
3. Vorhalt, Vorlauf, Vorzündung;
4. Steigung (einer Schraubenlinie u. dgl.).

Die mit „Blei" verwandten Bedeutungen sind: Plombe, Mine, Bleistiftmine, Lot, Senkblei.

Weitere Verbindungen mit *lead*: *lead-in groove* „Einlaufrille", *leader* „Vorspannband" (in der Magnettontechnik).

phosphor – phosphorus

In Großbritannien sowie in den USA wird der „Leuchtstoff" in Bildwiedergabe- und allen übrigen Kathodenstrahlröhren *phosphor* genannt, obwohl der Leuchtschirm keinen „Phosphor" (*phosphorus*, chemisches Element P) enthält. Mit *phosphorescence* grenzt man denjenigen Anteil der Lumineszenz ab, der nach der Erregung andauert, während *fluorescence* (ungefähr) den Anteil während der Erregung bezeichnet. Angesichts der schwierigen Auflösung zwischen beiden Anteilen zieht man im Deutschen die Benennungen *Postlumineszenz* bzw. *Kolumineszenz* vor.

technology – technique – engineering

Die englische Fachsprache verwendet die Benennung *technology* in einem viel breiteren Sinne als man unter „Technologie" im Deutschen versteht (Verfahrenstechnik, Werkstoffkunde, Lehre von der Gewinnung und Verarbeitung von Roh- und Werkstoffen; im engeren Sinne auch Verfahren und Methodenlehre eines einzelnen Ingenieurgebietes oder eines bestimmten Fertigungsablaufes).

In den meisten Fällen kann *technology* direkt mit „Technik" übertragen werden. Z. B. der technische Fortschritt heißt eindeutig *technological progress*, der technische Rückstand *technological gap*. Das Wort *technique* wird viel seltener benutzt als *technology* und bezeichnet eher „die Art, wie man etwas tut", z. B. unter *measuring technique* wird vielmehr die Art des Messens verstanden, als die technische Meßausrüstung, daher „Meßverfahren".

Auch Benennungen der technischen Gebiete, die im Deutschen mit „-technik" enden, werden im Englischen nur selten *technique* bezeichnet. „Elektrotechnik" heißt *electrical engineering*, und Wörter *electrotechnique, electrotechnical* sind im Englischen fast unbekannt. Das Adjektiv *electrotechnical* wird eigentlich nur im Titel der *IEC* (*International Electrotechnical Commission*) gebraucht.

wire gauge
  Der Ausdruck *wire gauge* /GB/, /USA/ oder *wire gage* /USA/ bezeichnet einerseits die Lehre zum Messen des Drahtdurchmessers, andererseits ein Maß für den Drahtdurchmesser. Die wichtigsten Lehrensysteme sind:
  – *British Standard Wire Gauge* (*BSWG*), auch *National British Standard* (*NBS*) *Wire Gauge* genannt,
  – *Birmingham Wire Gauge* (*BWG*) für Fernsprechleitungen (veraltet),
  – *American Wire Gage* (*AWG*), früher *Brown and Sharpe Wire Gage* (*B&SWG*).

  Britische und amerikanische Maßsysteme für Drähte sind absteigend (je höher die Zahl, desto dünner der Draht).

  Außerdem werden die Drahtdurchmesser auch in *mil* (s. S. 45) angegeben.

4. Manufacture of Electrical Equipment (Herstellung elektrischer Geräte)

4.1. Electrical Parts
(Elektrische Bauteile)

The three basic types of *parts* (also known as *components* or *component parts*) are *resistors*, *capacitors* and *inductors* (mainly *coils*).

The two main types of resistors are *carbon-composition resistors* and *wire-wound resistors*. *Variable resistors* are termed *rheostats*, while *potentiometers* are *voltage dividers* with a variable metal *arm* (or *slider*). Both the *fixed resistors* and variable resistors have certain *wattage ratings*, larger resistors being able to *dissipate* more power. Special types of resistors include *deposited-film resistors*, high-voltage resistors, insulated wire-wound resistors, *bobbin wire-wound resistors*, *cermet resistors* as well as *varistors* and *thermistors*.

Capacitors may be divided into two major groups – fixed and variable. *Fixed capacitors* may be classified according to the type of material used as the dielectric, such as *paper capacitors*, *oil capacitors*, *mica capacitors*, ceramic capacitors and electrolytic capacitors. The capacitance of variable capacitors is varied by rotating a shaft to make *rotor plates* (usually *live*) *mesh* with *stator plates* (usually grounded).

A coil consists of a number of *turns* of wire wound on a *coil form*. The *winding* may be *spool-wound* or *former-wound*, *cross-wound* or *scramble-wound* (*random-wound*). According to their circuit function, coils may be *coreless* or provided with *laminated*, *dust* or *ferrite cores*.

A combination of two coils wound over each other is a *transformer*. The *primary winding* and the *secondary winding* are magnetically coupled for transferring energy. Some of the transformers used in radio circuits are *air-core* types, but all the other types are wound on laminated cores. A transformer with an *interrupter* in its *d.c. supplied* primary circuit is called an *induction coil*. It is used for obtaining a high, *intermittent voltage* from a low-voltage source, for example in automobile *ignition systems*.

In addition to the three basic types of parts a great variety of other circuit elements are used in the manufacture of electrical equipment, such as *switches*, *relays*, *connectors*, *fuses*, *plugs* and *jacks*, etc. (Transistors and other *semiconductor devices*, including *integrated circuits*, are discussed in detail in Chapter 11, while various types of *vacuum tubes* are described in Chapter 12 of this book.)

air core	Luftkern *m*
arm	Arm *m*
bobbin wire-wound resistor	spulengewickelter Widerstand *m*

capacitor* IEC	Kondensator *m* DIN, IEC
carbon-composition resistor	Kohlewiderstand *m*
cermet resistor	Cermetwiderstand *m*
coil	Spule *f*
coil form	Spulenkörper *m*
component; component part ANSI	Bauelement *n* DIN; Element *n* NTG; Bauteil *n*
connector	Konnektor *m*; Steckverbindung *f*
coreless	kernlos
cross-wound	kreuzgewickelt
d.c. supplied	gleichstromgespeist
deposited-film resistor	Schichtwiderstand *m*
to dissipate	die Energie abführen
dust core	Staubkern *m*
ferrite core	Ferritkern *m*
fixed capacitor	Festkondensator *m*
fixed resistor	Festwiderstand *m*
former-wound	auf einem Spulenkörper gewickelt
fuse	Sicherung *f*
ignition system	Zündsystem *n*
induction coil	Induktionsspule *f*
inductor	Spule *f*
integrated circuit	integrierte (Mikro)schaltung *f*
intermittent voltage	unterbrochene Spannung *f*
interrupter	Unterbrecher *m*
jack	Buchse *f*; Klinke *f*
laminated core	Blechkern *m*
live*	unter Spannung stehend VDE
to mesh	(hier:) ineinandergreifen
mica capacitor	Glimmerkondensator *m*
oil capacitor	Ölkondensator *m*
paper capacitor	Papierkondensator *m* VDE
part ANSI	Bauelement *n* DIN; Element *n* NTG; Bauteil *n*
plug	Stecker *m*; Stöpsel *m*
potentiometer	Potentiometer *n*
primary winding	Primärwicklung *f*
random-wound	wild gewickelt
relay	Relais *n*
resistor	Widerstand *m*
rheostat	Rheostat *m*; Regelwiderstand *m*
rotor plates	Rotorplatten *fpl*
scramble-wound	wild gewickelt
secondary winding	Sekundärwicklung *f*
semiconductor device	Halbleiterbauelement *n*
slider	Schleifer *m*
spool-wound	spulengewickelt
stator plates	Statorplatten *f pl*
switch CEE, IEC	Schalter *m* DIN, VDE
thermistor	Heißleiter *m*
transformer	Transformator *m* DIN, VDE
turn	Windung *f*

vacuum tube	Elektronenröhre *f*
variable resistor	veränderbarer Widerstand *m*
varistor	Varistor *m*
voltage divider	Spannungsteiler *m*
wattage rating	Verlustleistung *f*
winding	Wicklung *f* VDE
wire-wound resistor	Drahtwiderstand *m*

A selection of related terms — **Ausgewählte verwandte Benennungen**

capacitor IEC	Kondensator *m* VDE
encased ~	eingekapselter ~
metallized paper ~	MP-~ VDE
organic film ~	Kunststoff-~ VDE
padder ~ ; padding ~	Padding~ ; Reihenabgleich~
self-healing ~	selbstheilender (elektrolytischer) ~
tantalytic ~	Tantal-~ DIN
trimmer ~	Trimmer~
coil	Spule *f*
bifilar ~	bifilargewickelte ~
honeycomb ~	Waben~
plug-in ~	Aufsteck~
printed ~	gedruckte ~
tapped ~	angezapfte ~
toroidal ~	Ring~
key	Hebelschalter *m*
powder core	Staubkern *m*
silicone* insulation	Silikonisolierung *f*
spare part	Ersatzteil *n*
switch	Schalter *m* DIN, VDE
commutating ~	Um~
detented ~	Einrast~
interlocked ~ es	verriegelte ~ *mpl*
mercury ~	Quecksilber~
micro-~ (Warenzeichen)	Mikro~
push-button ~	Druckknopf~ , Tasten~
rocker ~	Schwing~
rotary ~	Dreh~
slide ~	Schiebe~ ; Schleif~
snap-action ~	Schnapp~
stacked ~	~ mit Stapelkontakten
toggle-~	Kipp~
transformer	Transformator *m*
current ~	Stromwandler *m*
shell-type ~	Mantel~
step-down ~	Abwärts~
step-up ~	Aufwärts~

4.2. Construction of Equipment
(Gerätebau)

The construction of electrical equipment consists, mainly, in *assembling* electrical parts into *subassemblies* and/or *assemblies*. Collections of subassemblies and assemblies are then packaged together into *units*. The parts to be assembled according to a given *scheme* are connected by means of wires or cables using *splices, terminals* or *multipole connectors*.

High quality workmanship and materials must be employed in the *wiring* of *apparatus* to ensure reliable electrical contact, physical strength and insulation. Connections are made by *soldering* wires to *terminal lugs*, but *solderless joints* are often used for preventing *cold solder joints*, or burned insulation. The final step in completing a splice or joint may be the placing of *electrical tape*, or *friction tape*, over the bare wire.

In electronic and radio equipment construction, *printed circuits* have come into widespread use because of the many advantages associated with them. The decrease in cost when using *PC's* arises from the reduction in assembly labour, in necessary *testing* and *alignment* and in the low number of *rejects*. Another advantage is the application of *dip-soldering* techniques.

alignment	Einstellen *n*
apparatus IEC	Gerät *n* VDE
to assemble	zusammenbauen
assembly ANSI	Gruppe *f*
cold solder joint	Kaltlötstelle *f*
dip soldering	Tauchlötung *f*
electrical tape	Isolierband *n*; Klebeband *n*
friction tape	Isolierband *n*
joint	Verbindung *f*
multipole connector	Mehrpol-Steckverbindung *f*
PC; printed circuit	gedruckte Schaltung *f*
reject	Ausschuß *m*
scheme*	Anordnung *f*
to solder	(weich)löten
solderless	lötfrei
splice	Spleiße *f*
subassembly ANSI	Untergruppe *f*
terminal CEE, IEC	Klemme *f* VDE
terminal lug	Lötfahne *f*; Lötöse *f*
testing	Prüfen *n*
unit ANSI	Einheit *f*; Gerät *n*
wiring CEE, IEC	Verdrahtung *f* VDE

A selection of related terms	**Ausgewählte verwandte Benennungen**
britannia joint	Wickellötstelle *f*
frame; bay	Rahmen *m*
lead	Leiter *m*; Zuleitung *f*
pin	Sockelstift *m*
rack	Gestell *n*; Rahmen *m*
rack-and-panel type connector	Steckkontaktleiste *f*
rack-mount	Gestelleinschub *m*
soldering gun	Lötpistole *f*
soldering iron	Lötkolben *m*
soldering tag	Lötfahne *f*
terminal strip	Klemmenstreifen *m*

3. Manufacture of Plated Wire Memories
(Herstellung von Speichermatrizen aus metallbeschichtetem Draht)

Plated wire is a *proved memory technology* for both commercial applications and as a special-purpose military and aerospace memory (used in the Poseidon, Titan 111 and Apollo programmes). Plated wire for memory application consists of a fine beryllium-copper *substrate wire* onto which has been *electroplated* first a layer of copper and then a layer of a magnetic nickel-iron *alloy*. The *memory array* consists of an assembly of plated wires and *word line conductors*, the latter being a printed circuit *bonded* around a *moulded, grooved plastic carrier* to form a regular *tunnel structure* into which the plated wires are inserted. The tunnel-structure *package* is bonded to both sides of a rigid *baseboard* to make the basic *memory plane*. Memory planes are *stacked* by making interconnections between plated wires using *flexible wiring*.

Beryllium-copper wire (0·02in diameter) is first *etched* and then *drawn* using special techniques down to 0·006in diameter. Further cleaning and *heat treatment* is necessary before the wire goes into the *plating line*. Wire is pulled *at constant rate* through a *cleaning stage* and then the copper and nickel-iron plating stages. After plating, the wire is heat treated to stabilize the magnetic layer so that it maintains its *memory properties* for the *lifetime* of practical applications. Detailed *acceptable quality level* tests are then *run* on samples to check the *memory performance*. Accepted *batches* are 100 per cent tested and *localized defective areas* are cut out. The plated wires with their ends *presolder dipped* are *threaded* through the *tunnels* by simple hand methods.

The *transistor matrix wiring*, a 2-layer circuit using *plated through holes*, is also bonded to the baseboard. Transistor packs are mounted on the board to link the two circuits and complete the *plane* manufacture. The planes are then stacked and, after further d.c. tests, are *subjected* to *memory pulse tests* to identify defective *memory bits*.
(From: Component Technology, October 1970)

acceptable quality level EOQC; AQL	Annahmegrenze f EOQC; annehmbare Qualitätslage f EOQC
alloy	Legierung f
at constant rate	mit konstanter Geschwindigkeit
baseboard	Grundplatte f
batch EOQC	Charge f EOQC
to bond	kleben
cleaning stage	(hier:) Entfettungsanlage f
defective area	fehlerhafte Stelle f
to draw	ziehen
to electroplate	galvanisieren
to etch	ätzen
flexible wiring	biegsame (gedruckte) Verdrahtung f
to groove	mit Nuten versehen
heat treatment	Wärmebehandlung f
lifetime	Lebensdauer f
localized	(hier:) aufgefunden
memory array	Speichermatrix f
memory bit	(hier:) Speicherelement n
memory performance	Speicherfähigkeit f
memory plane	Speicherebene f
memory properties	Speichereigenschaften f
memory pulse test	Impulsprüfung f des Speichers
memory technology	Speichertechnik f
to mould	pressen
package	(hier:) Untergruppe f
plane	(hier:) Speicherplatte f
plastic carrier	(hier:) Kunststoff-Trägerplatte f

plated through hole	durchplattiertes Loch *n*
plated wire	galvanisierter Draht *m*
plating line	Plattierstraße *f*; Galvanisierstraße *f*
presolder dipped	von vorn herein im Tauchbad verzinnt
proved	erprobt
to run (tests)	(hier:) (Prüfungen) unternehmen
to stack	stapeln
to subject	unterwerfen
substrate wire	(hier:) Trägerdraht *m*
to thread	einfädeln
transistor matrix wiring	Verdrahtung *f* der Transistormatrix
tunnel	(hier:) Nut *f*
tunnel structure	tunnelartiges Gefüge *n*
word line conductor	Wortleitung *f*

Reliability and Quality Control
(Zuverlässigkeit und Qualitätssteuerung)

Reliability may be understood as the *probability* that a component, equipment or system will statisfactorily perform under given circumstances, such as *environmental conditions*, limitations as to *operating time*, and *frequency* and thoroughness of *maintenance*.

A *failure* is the inability of a component or equipment to carry out its specified function. *Sudden failures* are those which occur so quickly that they cannot be *predicted*, while *degradation failures* develop in such a manner that the performance gradually exceeds the permissible tolerance. The *failure density* gives the probability that a component *fails* at a certain time.

Various *reliability levels* are used for indicating the relation between the number of components in an equipment and the *mean time between failure* (*m.t.b.f.*) for the equipment as a whole.

Quality control engineering may be defined as the application of specialized managerial, scientific and technological *know-how* to achieve the desired quality at minimum costs by *inspecting* a small portion of the products, and *estimating* the over-all quality of the products to determine what, if any, changes must be made to achieve the quality level corresponding to the *state of the art*.

degradation failure	Driftausfall *m*
environmental condition	Umgebungsbedingung *f*
to estimate EOQC	abschätzen EOQC
to fail	ausfallen
failure	Ausfall *m*
failure density	Ausfallhäufigkeitsdichte *f*; Ausfalldichte *f*
frequency*	(hier:) Häufigkeit *f*
to inspect EOQC	prüfen EOQC
know-how	Know-how *n*; „wissen wie"
maintenance EOQC	Wartung *f* EOQC
mean time between failure; m.t.b.f. EOQC	mittlerer Ausfallabstand *m* EOQC
operating time	Betriebszeit *f*
to predict	vorhersagen
probability EOQC	Wahrscheinlichkeit *f* EOQC

quality control EOQC	Qualitätssteuerung *f* EOQC
quality control engineering EOQC	Qualitätssteuertechnik *f*
reliability EOQC	Zuverlässigkeit *f* NTG, EOQC
reliability level	Zuverlässigkeitsniveau *n*
state of the art*	Stand der Technik
sudden failure	Sprungausfall *m*

A selection of related terms — **Ausgewählte verwandte Benennungen**

Reliability — **Zuverlässigkeit** *f*

change	Änderung *f*
restorable ~	rückführbare ~
reversible ~	umkehrbare ~
defect	Fehler *f*
derating	Unterlastung *f*
deviation	Abweichung *f*
failure	Ausfall *m*
secondary ~	Folge~
blackout; catastrophic ~	Gesamt~
partial ~	Teil~
degradation ~	Drift~
random ~	Zufalls~
systematic ~	systematischer ~
wearout ~	Verschleiß~
failure rate	Ausfallrate *f*
full operating time	Vollbetriebszeit *f*
life	Lebensdauer *f*
malfunction	Störung *f*
scatter; dispersion	Streuung *f*
service life; utility life	Brauchbarkeitsdauer *f*
survivals	Bestand *m* (von Betrachtungseinheiten)
stress	Beanspruchung *f*

Quality Control EOQC — **Qualitätssteuerung** *f* EOQC

average outgoing quality (AOQ) EOQC	Durchschlupf *m* (fehlerhafte Stücke ersetzt) EOQC
average outgoing quality limit (AOQL) EOQC	größter Durchschlupf *m* EOQC
bias EOQC	Verzerrung *f* (systematischer Fehler) EOQC
biased sample EOQC	verzerrte Stichprobe *f* EOQC
bulk sampling EOQC	Stichprobenentnahme *f* aus Massengütern EOQC
confidence coefficient; confidence level EOQC	Vertrauenskoeffizient *m*; Aussagewahrscheinlichkeit *f*
confidence interval EOQC	Vertrauensbereich *m* EOQC
quality engineering EOQC	Qualitätstechnik *f* EOQC
sample EOQC	Stichprobe *f* EOQC
to sample EOQC	Stichprobe entnehmen EOQC
sampling interval EOQC	Entnahmeabstand *m*; Auswahlabstand *m* EOQC
test piece EOQC	Prüfling *m* EOQC

Fachsprachliche Anmerkungen

art
electrical art ist eine etwas gehobene Bezeichnung für Elektrizitätslehre, da im Wort *art* immer noch der ursprüngliche Sinn, nämlich „Kunst" im Sinne von Wissen, Kenntnis, Wissenschaft, Geschicklichkeit und Fertigkeit zu spüren ist. *state of the art* ist eine besonders im Arbeitsschutz und Patentwesen sehr oft gebrauchte stehende Wortgruppe, die genau durch „Stand der Technik" übersetzt werden kann, ebenso wie *rules of the art* durch „Regeln der Technik" wiedergegeben werden.
In den Vereinigten Staaten wird die Wortgruppe *Science and Arts* für „Theorie und Praxis" gebraucht.

capacitor – condenser
Im neuzeitlichen elektrotechnischen Fachschrifttum gibt es nur eine einzige Benennung für den (elektrischen) Kondensator – nämlich *capacitor*. Nur in Ausnahmefällen findet man noch die Bezeichnung *condenser* in Verbindungen wie *condenser microphone* IEC „elektrostatisches Mikrophon" oder *condenser bushing* „Hochspannungsdurchführungshülse".
Sehr bemerkenswert – da offensichtlich historisch begründet – ist die Bedeutung von *synchronous condenser* ANSI, BS, IEC „synchroner Phasenschieber"; es handelt sich also um eine umlaufende Maschine, die ohne Last läuft und nur zum Verbessern des Leistungsfaktors benutzt wird. Ein echter Kondensator zur Verbesserung des Leistungsfaktors (cos φ) heißt dagegen *static condenser*, obwohl diese Benennung weitschweifig zu sein scheint.
Auf den verwandten Gebieten begegnet man allerdings dem Wort *condenser* öfter, z. B. in der Bedeutung von „Dampfkondensator" oder „optischer Kondensor". Siehe auch Fachsprachliche Anmerkung zu *resistor* im Kap. 1.

family of -ability terms in reliability terminology (generally expressing reliability characteristics)
Im folgenden werden englische Benennungen auf dem Gebiete der Zuverlässigkeit, die mit **-ability** enden und allgemein Zuverlässigkeitskenngrößen bezeichnen, zusammengestellt:

EOQC, ANSI	EOQC, DIN
availability	Verfügbarkeit
capability	Leistungsfähigkeit
dependability	Systemstabilität; Anpassungsfähigkeit
maintainability	Wartbarkeit; Unterhaltbarkeit
probability	Wahrscheinlichkeit
reliability	Zuverlässigkeit
repeatability	Wiederholbarkeit
restorability	Instandsetzbarkeit

frequency
Im allgemeinen Sprachgebrauch ist *frequency* IEC (Frequenz DIN, IEC, UIP) die Zahl, mit der sich ein periodischer Vorgang in einer Zeiteinheit wiederholt.
Neben dieser naheliegenden Bedeutung sollte man aber nicht vergessen, daß *frequency* auch „Häufigkeit" oder „Rate" bedeuten kann, z. B. *frequency distribution* (Häufigkeitsverteilung) in der Statistik.

live – hot

live IEC oder seltener *alive* BS, IEEE, wörtlich „lebendig", bezeichnet in der Elektrotechnik den elektrischen Zustand eines Leiters, der auf einem Potential gegen Erde liegt. Bisweilen wird aber das Wort *live* oder *alive* mißdeutend im engeren Sinne statt *current-carrying* („stromdurchflossen") benutzt. Da viele Wörterbücher nur die letztgenannte Bedeutung angeben, weisen wir ausdrücklich darauf hin, daß sich die Bezeichnung *live* auch auf Fälle erstreckt, wo kein Strom zustande kommt. Daher ist *live* mit „unter Spannung stehend" VDE zu übersetzen.

Im englischen Fachjargon gibt es übrigens das Wort *hot* mit demselben Sinn wie *live*. Im deutschen Funkjargon bedeutet „heißes Ende" der HF-Spannung „führender Anschluß einer HF-Spule".

Außerdem bezeichnet *hot* – neben „warm" oder „heiß" – auch andere gefährliche Zustände, namentlich Radioaktivität (*hot laboratory*, *hot material*), sowie Zustände der Betriebsbereitschaft (*hot reserve*, *hot line*).

Auch *live* wird in vielen anderen übertragenen Bedeutungen benutzt: *a live transmission* ist eine direkte Rundfunksendung (auch neuerdings Live-Sendung genannt), *live steam* ist „Frischdampf", *live load* „bewegliche Belastung".

scheme – schedule – plan – project

Das Wort *scheme* kommt in vielen technischen und wissenschaftlichen Texten im Sinne „Plan", „Methode", „Anordnung", „Einteilung", nie aber im Sinne von „Schema" (= *diagram*, z. B. „Schaltschema" *wiring diagram*) vor. Oft kann *scheme* auch „Prinzip", „Grundkonzeption", „System" bedeuten, z. B. *four schemes of energy conversion* „vier Enegieumsetzungsprinzipien", *Certification Scheme* „Zulassungsverfahren der CEE".

schedule wird öfter gebraucht und bedeutet meistens eine festgelegte Reihenfolge, ein Verzeichnis (manchmal zahlenmäßig ausgedrücktes), Fahrplan (*train schedule*) u. ä. Dem weiteren Synonym *plan* wird ziemlich selten begegnet. Das Wort *project* entspricht ungefähr dem deutschen „Projekt".

silicon – silicone

Es soll hier auf eine unauffällige, aber wichtige Einzelheit in der Rechtschreibung der Wörter *silicon* und *silicone* hingewiesen werden: *silicon* (ohne „*e*") bezeichnet das chemische Element Silizium, z. B. in *silicon rectifier* „Siliziumgleichrichter" oder *silicon steel* „Siliziumstahl", während *silicone* (mit „*e*") als Oberbegriff für Silikone, d. h. organische Silizium-Verbindungen dient (z. B. *silicone oil* „Silikonöl").

In Verbindung mit dem Thema „Silizium" ist noch die Benennung *silica* erwähnenswert; sie bezeichnet die Verbindung „Siliziumdioxid".

5. Electrical Measurements
(Elektrische Messungen)

5.1. Fundamentals of Measurement Techniques
(Grundlagen der Meßtechnik)

Electrical measurements may be divided, in general, into *high precision measurements*, *laboratory measurements*, and *industrial measurements*. Analogously, *measurement instruments* may be divided into *standard instruments* of high *precision*, used particularly for *calibration* of other instruments, *substandard instruments*, precision instruments and industrial instruments for general use.

Certain general precautions should be observed in electrical measurements for avoiding *errors*. The probable *limits of accuracy* of the *standards*, instruments and methods should be known, and one measurement should not be relied upon but several *readings* should be taken.

The error of an *indicated value* is the difference between the indicated value and the *true value* of the *measurand* (the physical quantity to be measured). The *correction* of an error has the same numerical value as the error of the indicated value, but the opposite sign.

The accuracy of an indicated or *recorded value* is expressed by the ratio of the error of the indicated value to the true value, usually expressed in per cent. The *accuracy rating* of an instrument (a *meter*) designates the *accuracy classification* of the instrument. It is given as the limit, usually expressed as a percentage of *full-scale value*, which errors will not exceed when the instrument is used under *reference conditions*.

accuracy classification; class of accuracy IEC	Genauigkeitsklasse *f* VDE, IEC
accuracy rating; rating IEC	Meßbereichendwert *m* VDE, IEC; Meßbereichnennwert *m* IEC
calibration IEC	Eichung *f* IEC; Einmessen *n* IEC; Kalibrieren *n*
correction IEC	Korrektion *f* DIN, VDE; Korrektur *f* IEC
error	Fehler *m*
full-scale value	Endwert *m* des Meßbereichs
high precision measurement	Präzisionsmessung *f*
indicated value	Meßwert *m* DIN
industrial measurement	Betriebsmessung *f*
laboratory measurement	Labormessung *f*; Prüffeldmessung *f*
limits of accuracy	(hier:) Fehlergrenzen *fpl*
measurand	Meßgröße *f* DIN; Meßgegenstand *m* DIN
measurement instrument	Meßgerät *n* DIN, VDE; Meßinstrument *n*; Meßeinrichtung *f*
meter	Meßgerät *n*; Meßeinrichtung *f*
precision	Genauigkeit *f*
reading	Anzeige *f* DIN
recorded value	registrierter Wert *m*
reference conditions	Bezugsbedingungen *fpl*
scale IEC	Skale *f* DIN; Skalenteilung *f* IEC
standard	Normal *n*; Eichmaß *n*
standard instrument IEC	Laboratoriumsinstrument *n* höchster Genauigkeit IEC
substandard instrument IEC	Laboratoriumsinstrument *n* geringerer Genauigkeit IEC
true value	richtiger Wert *m*
A selection of related terms	**Ausgewählte verwandte Benennungen**
adjusting	Justieren *n*; Abgleichen *n*
deflection IEC	Ausschlag *m* IEC
effective range IEC	Meßbereich *m* DIN, VDE, IEC

error	Fehler *m*
absolute ~ IEC	absoluter ~ IEC
accidental ~ ; random ~	zufälliger ~
systematic ~	systematischer ~
relative ~ IEC	relativer ~ IEC
measurement method	Meßverfahren *n*
ballistic ~ ~	ballistisches ~
bridge ~ ~	Brücken~
comparison ~ ~	Vergleichs~
substitution ~ ~	Substitutions~
integration ~ ~	integrierendes ~
digital ~ ~	digitales ~
analogue ~ ~	analoges ~
scale factor	Skalenkonstante *f* DIN; Skalenfaktor *m* DIN; Konstante *f* VDE
sensitivity (absolute) IEC	Empfindlichkeit *f* DIN, IEC
telemetering IEC	Fernmessung *f* (als Tätigkeit)
telemetry	Fernmessung *f*, Telemetrie *f* (als Fachgebiet)

5.2. D.C. Measurements
(Gleichstrommessungen)

A direct current instrument is a *measuring device* using electromagnetic means to deflect a *pointer* over a calibrated scale. In the *permanent-magnet moving-coil instrument* a coil of wire, to which the pointer is attached, is *pivoted* between the poles of a permanent magnet. When current flows through the coil it produces a magnetic field that *interacts with* that of the magnet to cause the coil to turn (Fig. 5-1). A less expensive type of instrument is the *permanent-magnet moving-iron instrument*.

There are a great many types of d.c. meters. The most common are galvanometers, *ammeters*, *voltmeters*, *wattmeters*, and ohmmeters of various designs. Some of them are *suppressed zero* instruments or *zero-center* instruments, and often *shunts* and series resistors, called *multipliers*, are connected in parallel and in series, respectively, with the instrument proper. For measuring high resistance values, an instrument called a *megger* (megohmmeter) and incorporating a *hand-driven generator* is used. A *multimeter* is a multipurpose instrument that can measure various ranges of resistance, voltage or current.

A special type of meter based on the thermal effect of electric current is the *hot-wire instrument* with a *square-law scale* (Fig. 5-2).

ammeter IEC	Strommesser *m* DIN, IEC; Amperemeter *n* IEC
direct current instrument	Gleichstrom-Meßgerät *n*
hand-driven generator	Kurbelinduktor *m*
hot-wire instrument IEC	Hitzdrahtinstrument *n* VDE, IEC
to interact with	aufeinander wirken
measuring device IEC	Meßeinrichtung *f* IEC; Meßanordnung *f* IEC
megger	Isolationsmeßgerät *n*
multimeter	Vielfachmeßgerät *n*
multiplier	(hier allgemein:) Meßwiderstand *m*
permanent-magnet moving-coil instrument IEC	Drehspulinstrument *n* DIN, VDE, IEC

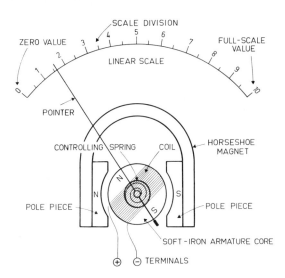

*Fig. 5–1. Moving-coil instrument.
(Drehspulinstrument.)*

coil	*Spule*
controlling spring	*Feder; Rückstellfeder*
full-scale value	*Endwert des Meßbereichs*
horseshoe magnet	*Hufeisenmagnet*
linear scale	*lineare Skale*
N – North	*Nordpol*
pointer	*Zeiger*
pole piece	*Polstück*
S – South	*Südpol*
scale division	*Skalenteilung*
soft-iron armature core	*Ankerkern aus Weicheisen*
terminal	*Klemme*
zero value	*Nullpunkt*

permanent-magnet moving-iron instrument IEC	Eisennadelinstrument *n* VDE, IEC; Dreheiseninstrument *n* mit Magnet IEC
pivoted	drehbar (an einem Zapfen) angeordnet
pointer IEC	Zeiger *m* IEC
shunt IEC	Nebenwiderstand *m* VDE, IEC; Shunt *m* IEC
square-law scale	quadratische Skalenteilung *f*
suppressed zero IEC	unterdrückter Nullpunkt *m* IEC
voltmeter IEC	Spannungsmesser *m* IEC; Voltmeter *n* IEC
wattmeter IEC	Leistungsmesser *m* IEC; Wattmeter *n* IEC
zero-center	Nullpunkt *m* in der Mitte der Skale

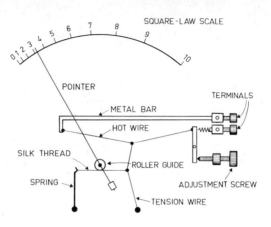

*Fig. 5–2. Hot-wire instrument.
(Hitzdrahtinstrument.)*

adjustment screw	Justierschraube
hot wire	Hitzdraht
metal bar	Metallsteg
pointer	Zeiger
roller guide	Rollenführung
silk thread	Seidenfaden
spring	Feder
square-law scale	quadratische Skalenteilung
tension wire	Spanndraht
terminals	Klemmen

5.3. A.C. Measurements
(Wechselstrommessungen)

Some of the simpler d.c. indicating instruments may be used for a.c. measurements as well, but special a.c. devices are intended for various purposes, such as *hook-on meters*, *instrument transformers*, *bridge meters*, *vibrating reed instruments*, or *power-factor meters*.

Two basic methods have been devised for adapting d.c. meters to a.c. measurements – the rectifier method for *converting* an alternating current to a *pulsating direct current*, and the *thermocouple* method, an indirect process that first converts the alternating current into heat and then uses this heat to generate a direct current. To avoid connecting instruments directly into *high-voltage lines*, *voltage transformers* and *current transformers* are used. Let us cite, as an example of a special a.c. instrument, the *watt-hour meter* (Fig. 5–3).

bridge meter; bridge IEC	Meßbrücke *f* IEC
to convert	umwandeln
current transformer IEC	Stromwandler *m* IEC
high-voltage line	Hochspannungsleitung *f*
hook-on meter; split-core type transformer IEC	Zangenwandler *m* IEC; Anlegewandler *m* IEC
instrument transformer IEC	Meßwandler *m* VDE, IEC

power-factor meter	Leistungsfaktormesser *m* IEC
pulsating* direct current	pulsierender Gleichstrom *m*
thermocouple IEC	Thermoelement *n* IEC
vibrating reed instrument IEC	Zungenfrequenzmesser *m* IEC
voltage transformer IEC	Spannungswandler *m* VDE, IEC
watt-hour meter IEC	Wattstundenzähler *m* IEC; Wirkverbrauchszähler *m* IEC

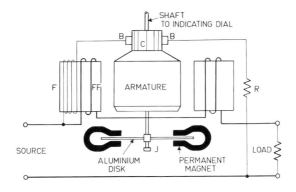

Fig. 5–3. Watt-hour meter.
(Wirkverbrauchszähler.)

aluminium disk	*Aluminiumscheibe*
armature	*Anker*
B – brushes	*Bürsten*
C – commutator	*Kommutator, Stromwender*
F, FF – coils	*Spulen*
J – jewel bearing	*(Edel)Steinlager*
load	*Last*
permanent magnet	*Dauermagnet*
R – resistor	*Widerstand*
shaft to indicating dial	*Welle (zu der Anzeigescheibe)*
source	*Stromquelle*

4. Radio-Frequency Measurements
(Hochfrequenz-Messungen)

At frequencies above a few kilohertz the common a.c. instruments become impractical because *stray capacitance* and other *second-order effects* make it difficult to maintain reasonable accuracy, and electronic instruments must be used. For r.f. current and voltage measurements *vacuum-tube voltmeters* (*VTVM*) or equivalent *transistor devices* are used, and signal generators serve for frequency measurements and *apparatus alignment*. The *standing-wave ratio* (*S.W.R.*) of *transmission lines* is measured by using *S.W.R. bridges*, and *field-strength meters*, *noise generators*, *grid-dip meters* and numerous other specific devices are used.

For giving a visual representation of *signal waveforms*, oscilloscopes and oscillographs are currently used in radio engineering. For observation of waveforms as functions of time it is necessary to generate and apply to the horizontal or x-axis *de-*

flection plates of a *cathode-ray tube* (see Fig. 12–4), *sweep voltages* or time-varying *saw-tooth voltages*. Any *recurring voltages* applied to the vertical or y-axis plates will then appear *plotted* on the *screen* of the C.R.T. as functions of time. The versatility of a *"scope"* can be greatly increased by adding *amplifiers* and *linear deflection circuits* to its *circuitry*.

amplifier	Verstärker *m* DIN
apparatus alignment	Abgleichen *n* von Geräten
cathode-ray tube	Kathodenstrahlröhre *f*; Elektronenstrahl-Oszillographenröhre *f* DIN
circuitry	Schaltanordnung *f*
C.R.T.; cathode-ray tube	Kathodenstrahlröhre *f*; Elektronenstrahl-Oszillographenröhre *f* DIN
deflection plate	Ablenkplatte *f*
field-strength meter	Feldstärkemeßgerät *n*
grid-dip meter	Grid-dip-Meter *n*
linear deflection circuit	linearer Ablenkkreis *m*
noise generator	Rauschgenerator *m*
to plot	aufzeichnen
recurring voltage	wiederkehrende Spannung *f*
saw-tooth voltage	Sägezahnspannung *f*
scope; oscilloscope IEC	Oszilloskop *n* IEC; Oszillograph *m*
screen	Schirm *m*
second-order effect	Effekt zweiter Ordnung
signal waveform	Form *f* von Signalwelle
standing-wave ratio; S.W.R.	Stehwellenverhältnis *n*
stray capacitance	Streukapazität *f*
sweep voltage	Kippspannung *f*
S.W.R. bridge	Welligkeitsmeßbrücke *f*
transistor device	Transistorgerät *n*
transmission line	Übertragungsleitung *f*
vacuum-tube voltmeter; VTVM	Röhrenvoltmeter *n*

A selection of related terms

Ausgewählte verwandte Benennungen

ampere-hour meter IEC	Amperestundenzähler *m* IEC
bearing	Lagerung *f*
chart	Registrierpapier *n*
deflecting torque	ablenkendes Drehmoment *n*
dial IEC	Skale *f* DIN, IEC; Skala *f* VDE
instrument	Instrument *n*
damped periodic ~ IEC	gedämpft schwingendes ~ IEC
D' Arsonval ~	Drehspul~ VDE, IEC
deflection ~	Zeiger~ IEC
detecting ~ IEC	Indikator *m* IEC
differential measuring ~	Differenzmesser *m*
direct-reading ~ IEC	direkt anzeigendes ~ IEC
flush-type ~	versenktes Gerät *n*
indicating ~ IEC	anzeigendes Meßgerät *n* IEC
integrating ~ IEC	integrierendes Meßgerät *n* IEC
mirror ~ IEC	Spiegel~ IEC
moving-iron ~ IEC; ferromagnetic ~	Dreheisen~ VDE, IEC
moving magnet ~ IEC	Drehmagnet~ VDE, IEC

moving-vane ~	Dreheisen~ VDE, IEC
pointer ~ IEC	Zeiger~ IEC
projected-scale ~ IEC	Projektionsskalen-~ IEC
recording ~ IEC	Schreiber *m* IEC; registrierendes Meßgerät *n* IEC
rotating field ~ IEC	Drehfeld~ IEC
shadow column ~ IEC	Schattenzeiger~ IEC
thermal ~ /GB/ IEC; electrothermal ~ /USA/ IEC	(elektro)thermisches ~ IEC
thermocouple ~ IEC	Thermoumformer~ IEC
instrument damping	Gerätdämpfung *f*
loop oscilloscope	Schleifenoszillograph *m*
peak voltmeter	Scheitelspannungsmesser *m* IEC
pivot IEC	Spitze *f* IEC; Zapfen *m* IEC
range selector; range multiplier	Meßbereichsschalter *m*
response time (of an instrument)	Einstellzeit *f*; Einstelldauer *f*
restoring torque IEC	Rückstellmoment *n* IEC; Richtmoment *n* IEC
sensing element*; sensor* IEEE	Meßumformer *m* DIN
thermopile IEC	Thermosäule *f* IEC
torsionless suspension	torsionslose Aufhängung *f*
var-hour meter IEC	Blindverbrauchszähler *m* IEC
varmeter IEC	Blindleistungsmesser *m* IEC; Varmeter *n* IEC

5. The SM 523 digital voltmeter
(Das digitale Voltmeter SM 523)

A new *programmable, systems orientated, digital voltmeter* is announced by Marconi Italiana, Genoa, Italy. The new *d.v.m.* is an accurate, linear, *all-solid-state instrument* measuring 0 to $\pm$ 1,200 volts D.C.

The SM 523 has an accuracy of 0.02% of full scale range. Voltage is indicated on a five-tube *in-line readout*, with fully automatic polarity indication and a decimal point coupled to the *range selector*. The indicator tubes are actuated by a *b.c.d. store*, whose *logic outputs* are also taken to a socket at the *rear* of the instrument for operation of a *printer* or other external digital equipment.

The SM 523 has programmable *sampling* and *ranging*, allowing *remote range selection* of its five voltage ranges. *Auto-ranging* is also available as an *optional facility*. Apart from its programmability and systems orientation, the principal features are its good zero stability, and its fast operation.

The high level of accuracy achieved is largely due to the *successive approximation technique* employed. With this technique, the input voltage to the instrument is compared with an internal voltage derived via a digital potentiometer from a *reference source*. The *difference potential* goes to a logic circuit which converts it into *command signals* that adjust the digital potentiometer for minimum difference potential. It is effectively the setting of the potentiometer that is indicated by the in-line readout as the measured voltage. A *built-in standard cell* is provided for calibrating the reference source.

In addition to its use for accurate measurement of voltage, the SM 523 can function as an analogue-to-digital converter. The b.c.d. output taken in conjunction with the facilities for programmable ranging and *triggering*, renders the instrument particularly suitable for use in automatic measurement or *control systems*.

The SM 523 is normally supplied as a *bench instrument*, but a *rack-mounting kit* is available as an optional accessory.
(From: Marconi Instruments Ltd. News Release, 1969)

all-solid state instrument	vollkommen in Festkörpertechnik ausgeführtes Gerät *n*
auto-ranging	automatische Meßbereichswahl *f*
b.c.d. store	BCD-Code-Speicher *m*
bench instrument	Tischgerät *n*
built-in standard cell	eingebaute Normalzelle *f*
command signal	Führungssignal *n*
control system	Regel- und Steuersystem *n*
difference potential	Potentialdifferenz *f*
digital voltmeter; d.v.m.	digitales Voltmeter *n*
in-line readout	direkte Anzeige *f*
logic output	Ausgabe *f* der Verknüpfungsglieder
optional facility	fakultative Einrichtung *f*
printer	Drucker *m*
programmable	programmierbar
rack-mounting kit	Bausatz *m* zum Einschub des Gerätes in ein Gestell; Gestelleinschub(bausatz) *m*
range selector	Bereichswähler *m*
ranging	Meßbereichumschaltung *f*
rear	(hier:) Hinterwand *f*
reference source	Bezugsspannungsquelle *f*
remote range selection	Fernwahl *f* des Meßbereiches
sampling	(hier:) Meßplatzumschaltung *f*
successive approximation technique	Iterationsverfahren *n*
systems orientated	systemorientiert
triggering	Auslösen *n*

Fachsprachliche Anmerkungen

cycles, amps, mhos, etc.
Im amerikanischen Fachschrifttum trifft man immer noch Bezeichnungen, Einheitennamen, Abkürzungen usw., die dem Stand der Normung nicht entsprechen. In vielen Zeitschriften, die für Leserkreise aus der Praxis bestimmt sind, wird statt in Hertz (*Hz*) die Frequenz in *cycles per second* (*c.p.s.*) oder gar nur in *cycles* (*c*), bzw. *kilocycles* und *megacycles* (*kcs* und *mcs*) angegeben.

Strom wird bisweilen immer noch in *amps* und Leitwert in *mhos* gemessen. Dazu kommt noch, daß die Vorsatzzeichen nicht eingehalten werden (siehe Anmerkung unter *kilo – mega – pico – micro*, Kap. 2).

impulse – pulse
Der Benennung *impulse* wird im englischen Fachschrifttum ziemlich selten begegnet. Sie wird in der Mechanik in demselben Sinne wie die deutsche Benennung „Impuls" (Zeitintegral der Kraft) benutzt. In der Elektrotechnik findet man sie nur im Sinne „Stoß" oder höchstens „Einzelimpuls" vor, wie z. B. in *impulse excitation* „Stoßanregung, Stoßerregung".

Sonst wird *impulse* durch sein Synonym *pulse* immer mehr verdrängt, und die britische terminologische Norm betrachtet sogar *impulse* als „veraltet". Die deutsche

Benennung „Impuls", die einen einmaligen, stoßartigen Vorgang endlicher Dauer bezeichnet, wird daher eindeutig mit *pulse* übersetzt.

Im Deutschen wird die Form „Puls" dagegen nur für eine periodische Folge von Impulsen gebraucht (siehe dazu *pulse train* im Kap. 14), und weiter in den Benennungen bzw. Abkürzungen der verschiedenen Pulscode-Modulationsarten, wo sich die englische Schreibweise wie PCM, PAM usw. fest eingebürgert hat (siehe auch *pulse modulation* im Kap. 14).

loop

Im elektrotechnischen Schrifttum kommt meistens *loop* im Sinne „Schleife, Schlinge" vor. In der Antennentechnik kann jedoch *loop* entweder „Rahmen" (Rahmenantenne, *loop antenna, frame antenna*) oder „Wellenbauch" (*antinode*) bedeuten.

mil – mill

Obwohl die Einführung des metrischen Systems (*metrication*) in Großbritannien weit fortgeschritten ist und in den USA wissenschaftliche Werke seit einigen Jahren überwiegend mit metrischen Einheiten abgefaßt werden, begegnen wir in den kommenden Jahren oft noch den Zolleinheiten.

Für kleine Längen wird noch häufig ein Tausendstel Zoll, *mil* (0,0254 mm) gebraucht. In dieser Einheit werden z. B. Schichtdicken angegeben. Der spezifische Widerstand wird manchmal auf *mil-foot* bezogen (Draht mit 1 *mil* Durchmesser und 1 *ft* Länge). Für Kupfer sind es z. B. 10,37 Ω.

Für Flächen wird oft neben *square mil* (Quadratmil) die Einheit *circular mil* (mit dem Kurzzeichen *cmil*) gebraucht. Dies ist die Fläche eines Kreises mit 1 *mil* Durchmesser und gleicht $\frac{\pi}{4} \cdot 10^{-6}$ in^2 (ein Quadratmillimeter gleicht 1974 *cmil*). Der Querschnitt eines Drahtes mit 1 *mil* Durchmesser ist also 1 *cmil*.

Die Einheit *mil* sollte nicht mit *mill* verwechselt werden; in *mills*, d. h. Tausendstel eines U.S.$ werden oft kleine Beträge angegeben, z. B. spezifische Kosten in der Energiewirtschaft.

Im Fachjargon wird statt *mil* die Einheit *thou* (lies [θau]) verwendet, die von *thousandth* (= $^1/_{1000}$) abgeleitet worden ist. *thou* bedeutet jedoch $^1/_{1000}$ im allgemeinen, nicht nur $^1/_{1000}$ Zoll.

propagation constant

Übertragungsmaß DIN, bzw. Übertragungskonstante DIN, Fortpflanzungsmaß DIN bzw. -konstante DIN. Im Deutschen wird streng darauf geachtet, ob die Übertragungsgrößen für die gesamte Leitung oder je Längeeinheit gelten: im ersten Falle spricht man vom „-maß", im zweiten von „-konstante". Ebenso muß man bei *attenuation* zwischen „Dämpfungsmaß" und „-konstante" und bei *phase constant* zwischen „Phasenmaß" und „-konstante" unterscheiden.

sensing elements

sensing elements, sensors IEEE, *measuring units* BS, IEC sind „Signalumformer" DIN oder „Meßumformer" DIN, die ein Eingangssignal – gegebenenfalls unter Verwendung einer Hilfsenergie – in ein damit zusammenhängendes Ausgangssignal umformen. Oft werden dafür auch folgende Ausdrücke benutzt: *primary element, detecting element, primary detector, measuring transmitter, transducer* und *pickup*. Die letzten zwei Benennungen werden überwiegend für Signalumformer für das Bearbeiten von Signalen auf dem Gebiete der Elektroakustik verwendet.

Die BS-Norm unterscheidet noch zwischen *detecting element*, das auf die Eingangsgröße anspricht, und *measuring element*, das auf das Signal vom *detecting element* anspricht und es umformt. *measuring unit* BS, IEC ist aus einem *measuring element* BS und einem *detecting element* BS, IEC zusammengesetzt.

Im folgenden wird versucht, die verschiedenen Signalumformerarten mit Rücksicht auf die in den Normen enthaltenen Definitionen und Erläuterungen systematisch gegenüberzustellen.

sensors IEEE, *measuring units* BS, IEC
Signalumformer DIN, Meßumformer DIN

		passive BS ohne Hilfsenergie	*active* BS, *self-operated* IEC mit Hilfsenergie
Mit gleicher Signal- und physikalischer Struktur am Eingang und Ausgang		*measuring element* BS Signalwandler DIN	Signalverstärker DIN
Mit unterschiedlicher	physikalischer Struktur am Eingang und Ausgang	*primary detector* *detecting element* BS Fühler DIN	
	Signal-Struktur	*transducer* (Meßgrößen)umformer	*signal converter* Signalumsetzer DIN Signalwandler IEC

Electric Power Engineering
(Starkstromtechnik)

Electric Power Production
(Erzeugung elektrischer Energie)

Power Plants
(Kraftwerke)

Electric power generation usually implies large-scale production in stationary *power plants* where *electric generators* convert energy from falling water (hydroelectric plants), or from coal, oil or natural gas (*fossil-fuel* plants, Fig. 6–1).

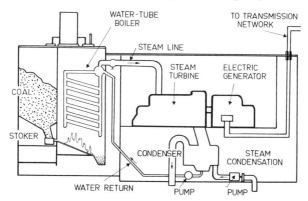

Fig. 6–1. Cross-section of a steam power plant.
(Schnitt durch ein Dampfkraftwerk.)

coal	Kohlen
condenser	(Dampf-)Kondensator
electric generator	Generator
pump	Pumpe
steam condensation	Dampfkondensation
steam line	Dampfleitung
steam turbine	Dampfturbine
stoker	Rostbeschicker
transmission network	Übertragungsnetz
water return	Rückfluß; Wasserrückführung
water-tube boiler	Wasserrohrkessel

The part of a power plant that transforms *energy* from the thermal, or the pressure form, to the mechanical form is called a *prime mover*. It is frequently an engine or turbine represented by such machines as water wheels, hydraulic turbines, *steam condensing turbines* or gas turbines, *internal combustion machines*, and *jet engines*. Among a great number of power-plant circuits and equipment those of major importance are the devices required for *generator protection* and *voltage regulation* and for synchronization of generators (by *matching* the instantaneous voltage of an *incoming source* to that of a *running source* and then connecting them together).

The latest type of *generating station* is the *atomic power plant* in which *nuclear fuels* are utilized in a *nuclear reactor*. These stations are in the development stage with a number of large-scale *fission reaction* plants in operation or under construction. *Fusion reaction* plants are still in the *R & D stage*.

atomic power plant	Kernkraftwerk *n* IEC; Atomkraftwerk IEC
electric generator	Generator *m* DIN, VDE
energy*	Energie *f*
fission reaction	Kernspaltung *f* DIN
fossil-fuel	fossiler Brennstoff *m*
fusion reaction	thermonukleare Reaktion *f*
generating station IEC	Kraftwerk *n* IEC
generator protection	Generatorschutz *m*
incoming source	(hier:) äußere Quelle *f*
internal combustion machine	Verbrennungsmotor *m* VDE
jet engine	Strahlmotor *m*
to match (to)	anpassen (an)
nuclear fuel	Kernbrennstoff *m* DIN
nuclear reactor	Kernreaktor *m* DIN
power* plant	Kraftwerk *n* IEC
prime mover	Kraftmaschine *f*; Primärmaschine *f*
R & D stage; research and development stage	Forschungs- und Entwicklungsstadium *n*
running source	(hier:) innere Quelle *f*
steam condensing turbine	Kondensationsdampfturbine *f*
voltage regulation	Spannungskonstanthaltung *f*

A selection of related terms

Ausgewählte verwandte Benennungen

Fossil-Fuel Plants

(Wärme)kraftwerke auf fossile Brennstoffe

boiler	Dampfkessel *m*
BTU; British Thermal Unit	BTU; britische Wärmeeinheit *f*
desuperheater	Dampfkühler *m*
economiser	Speisewasservorwärmer *m*
stoker	Rostbeschicker *m*
superheater	Überhitzer *m*

Hydroelectric Plants

Wasserkraftwerke

dam	Damm *m*, Staumauer *f*
draft tube	Ablaufkanal *m*
head	Gefälle *n*, Druckhöhe *f*
penstock	Zufuhrkanal *m*; Druckleitung *f*
spillway	Überlauf *m*

Power Plant Performance Figures

Kraftwerkkennwerte

absolute peak IEC	Jahresspitze *f* IEC
baseload	Tiefstlast *f*; Niedrigstlast *f*; Grundlast *f*
demand IEC	kurzzeitig gemittelte Belastung *f* IEC
factor	Faktor *m*
availability ~ IEC	Verfügbarkeits~ IEC
coincidence ~ IEC	Gleichzeitigkeits~ IEC
demand ~ IEC	Verbrauchs~ IEC
diversity ~ IEC	Verschiedenheits~ IEC

load ~ IEC	Belastungs~ (der kurzzeitig gemittelten Höchstlast) IEC
plant load ~ IEC; average load per cent IEC	Ausnutzungs~ eines Kraftwerkes IEC
installed capacity IEC; installed power IEC	installierte Leistung f IEC
load IEC	Last f IEC
load curve IEC	Belastungskurve f IEC; Ganglinie f IEC
maximum capacity	Höchstleistung f; Höchstlast f
peak load IEC	Belastungsspitze f IEC; Lastspitze f IEC
maximum demand IEC	kurzzeitig gemittelte Höchstlast f IEC
off-peak periods IEC	Schwachlastzeiten fpl IEC
output* IEC	(abgegebene) Leistung f IEC
overload IEC	Überlast f IEC
potential peak periods	Starklastzeiten fpl; Spitzenzeiten fpl
throughput*	durchgehende Leistung f

2. Electric Generators
(Elektrische Stromerzeuger)

Electric generators fall into two main groups, *alternating-current* (*a.c.*) and *direct-current* (*d.c.*). An elementary generator having a *stationary field* and a single, rotating, *armature coil* is shown in Fig. 6–2.

When the number of lines of magnetic flux in a generator is caused to change, an e.m.f. is generated in the coil, so that voltage may be induced in the *winding* by mechanically driving one member relative to the other. If the coil terminals were brought to a 2-segment *commutator* instead of *slip rings*, a *pulsating d.c. voltage* would appear at the *brushes* instead of a.c. voltage at the slip ring.

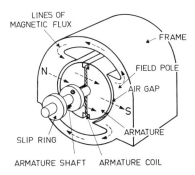

Fig. 6–2. Sketch of an elementary electric a.c. generator.
(Skizze eines einfachen Wechselstromgenerators.)

air gap	*Luftspalt*
armature	*Anker*
armature coil	*Ankerspule*
armature shaft	*Ankerwelle*
field pole	*Feldpol*
frame	*(hier:) Stator*
lines of magnetic flux	*Flußlinien (N, S)*
slip ring	*Schleifring*

The most common type of a.c. generator is the *synchronous generator*, sometimes termed an *alternator*. Another type is the induction generator, the speed of which varies somewhat with load for constant *output frequency*.

A d.c. generator is a *rotating electric machine* which delivers a *unidirectional current* rectified by a commutator mounted on the *rotor shaft*. D.c. generators are either *self excited generators* (*series generators*, *shunt generators*, and *compound generators*), or *separately excited generators*.

alternating-current generator; a.c. generator	Synchrongenerator *m*; Wechselstromgenerator *m* VDE
alternator	Wechselstromgenerator *m* VDE
armature coil	Ankerwicklung *f*
brush	Bürste *f*
commutator	Stromwender *m*; Kommutator *m*
compound generator	Doppelschlußgenerator *m*
direct-current generator; d.c. generator	Gleichstromgenerator *m* VDE
output frequency	Ausgangsfrequenz *f*
pulsating d.c. voltage	pulsierende Gleichspannung *f*
rotating electric machine	umlaufende elektrische Maschine *f*
rotor shaft	Rotorwelle *f*
self excited generator	selbsterregter Generator *m*
separately excited generator	fremderregter Generator *m*
series generator	Reihenschlußgenerator *m*
shunt generator	Nebenschlußgenerator *m*
slip ring	Schleifring *m*
stationary field	stationäres (magnetisches) Feld *n*
synchronous generator	Synchrongenerator *m*
unidirectional current	(hier:) Gleichstrom *m* DIN
winding	Wicklung *f* VDE

A selection of related terms	**Ausgewählte verwandte Benennungen**
booster	Zusatzmaschine *f*
condenser	Blindleistungsmaschine *f* VDE; Phasenschieber *m*
converter* BS, IEC; convertor* ANSI	Umformer *m* VDE
direct-current balancer	Ausgleichmaschine *f*
exciter	Erreger(maschine) *f*
generator	Generator *m*; Stromerzeuger(maschine) *f*
inductor ~	Induktormaschine *f*
multiple-current ~	Mehrstrom~
reluctance ~	Reaktions~
self-excited a.c. ~ with revolving armature	selbsterregter Wechselstrom~ mit umlaufender Kommutatorwicklung
homopolar machine	Unipolarmaschine *f*
induction generator	Asynchrongenerator *m*
inverter*	Wechselrichter *m* VDE

6.3. Chemical Sources of Electric Energy (Chemische Stromerzeuger)

A device which converts chemical energy into electric energy is called an electric battery. The term "battery" should by applied to a group of two or more *electric cells*; however in common usage it is also applied to a single cell.

Electric cells are conveniently treated in two groups: (1) *primary cells* which are designed to operate *irreversibly*, and *secondary cells* which are designed to operate reversibly. Their essential function is to store electric energy and for that reason they are also called *storage batteries*.

Primary cells are constructed so that only one continuous or *intermittent* discharge can be obtained. The construction of one type of a *dry cell* is shown in Fig. 6–3. The electrodes have the form of a carbon rod and a zinc container, the electrolyte may be an *acid*, *alkaline* or neutral *solution*.

Secondary cells are constructed so that they may be *recharged*. The most usual types of these cells are *lead-acid cells* and *alkaline cells*.

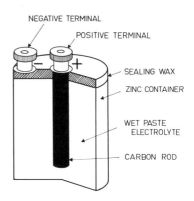

Fig. 6–3. Cross-sectional view of a dry cell.
(Schnitt durch eine Trockenzelle.)

carbon rod	Kohlestab
negative terminal	Minuspol
positive terminal	Pluspol
sealing wax	Verguβwachs; Wachsabdichtung
wet paste electrolyte	feuchte Elektrolytpaste
zinc container	Zinkbehälter

The main characteristics of electric cells and batteries are their *current capacity* (rated *in ampere-hours*), *working capacity* (rated in watt-hours), nominal voltage and current, and *internal resistance*.

acid solution	Säurelösung *f*
alkaline cell	alkalische Zelle *f*
alkaline solution	alkalische Lösung *f*
current capacity (in ampere-hours)	Kapazität *f* (in Amperestunden) VDE
dry cell IEC	Trockenzelle *f*; Trockenelement *n* IEC
electric cell	elektrische Zelle *f*
intermittent	zeitweilig unterbrochen, intermittierend
internal resistance	Innenwiderstand *m*
irreversibly	nicht umkehrbar; irreversibel
lead-acid cell	Blei-Akkumulator-Zelle *f*
primary cell* IEC	Primärzelle *f*; Primärelement *n* IEC
to recharge	neu aufladen; wiederladen
secondary cell* IEC	Sekundär-Zelle *f*; Sammlerbatterie *f* IEC; Akkumulatorenbatterie *f* IEC
storage battery*	Akkumulator *m* VDE
working capacity	Arbeitsvermögen *n*

A selection of related terms	**Ausgewählte verwandte Benennungen**
cell*	Zelle
gas-tight ~	gasdichte ~ VDE
open type ~	offene ~ VDE

charge IEC	Ladung *f* VDE, IEC
boost ~ IEC; quick ~	Schnell~ VDE, IEC
equalizing ~ IEC	Ausgleichs~ VDE, IEC
full ~	Voll~ VDE
initial ~	Inbetriebsetzungs~ VDE
over~	Überladen *n* VDE
trickle ~ IEC	Puffer~ IEC; Dauerladen *n* VDE
discharge IEC	Entladung *f* IEC
exhaustive ~	Tief~
self-~ (owing to local action) IEC	Selbst~ IEC; Selbstentladen *n* VDE
filler opening	Füllöffnung *f*; Einfüllöffnung *f*
filler plug	Füllpfropfen *m*; Einfüllstopfen *m*
forming IEC; formation	Formierung *f* IEC; Formieren *n* VDE
ironclad plate	Panzerplatte *f*
operation	Betrieb *m*
cycle ~ ; cycling ~	Lade-Entlade~ VDE; Zyklisieren *n*
floating ~	Puffer~ VDE
sealing compound	Vergußmasse *f*; Verschlußmasse *f*
separator IEC	Scheider *m* IEC; Separator *m* IEC
voltage	Spannung *f*
final discharge ~	Entladeschluß~ VDE
open-circuit ~ /USA/ IEC	Leerlauf~ ; Ruhe~

6.4. Novel Sources of Electric Energy
(Neue Quellen elektrischer Energie)

Each form of electric current generation is based on the conversion of stored energy into electric power. The fewer the conversion steps, the more efficient the process, and this is one of the reasons why the industry is developing new *power converters*.

The utilization of *solar energy* has presented a challenge for many years, and it is only lately that *solar cells* have been practically used on remote sites, in space vehicles, and the like.

Another promising method is the use of *fuel cells* in which reduction occurs at the cathode, *liberating* electrons to the external circuit. The fuel is *oxidized* at the anode, and an electrolyte completes the internal circuit.

The *thermoelectric method of generation* depends on the use of certain materials, such as oxides, sulfides, selenides, or tellurides of *transition metals*, capable of producing an electric current when heated. On the other hand, a thermionic converter is essentially a diode vacuum tube. Application of heat to the cathode *releases* electrons by *imparting* sufficient energy to some of them and enabling them to *surmount* the *work function* potential barrier.

A further possibility, the use of *magneto-hydrodynamic (MHD) generators*, is based on the generation of electric current directly from hot gases by passing a high temperature *gas stream* at high velocity through a strong magnetic field. The *lines of force* of the magnetic field *divert* free electrons into the external circuit.

to divert	ablenken
fuel cell	Brennstoffzelle *f*
gas stream	Gasstrom *m*
to impart	geben; „zuteilen"
to liberate	lösen; freisetzen
line of force	Feldlinie *f*

magneto-hydrodynamic generator; MHD generator	magnetohydrodynamischer Generator *m*; MHD-Generator *m*
to oxidize	oxidieren
power converter	Energiewandler *m*
to release	lösen
solar cell; solar battery	Sonnenzelle *f*; Sonnenbatterie *f*
solar energy	Sonnenenergie *f*
to surmount	überwinden
thermoelectric method of generation	thermoelektrische Stromerzeugung *f*
transition metal	Übergangsmetall *n*
work function	Austrittsarbeit *f*

Fachsprachliche Anmerkungen

cell – battery

Der Unterschied zwischen beiden Begriffen ist eindeutig: *galvanic cell* IEC „galvanisches Element" DIN IEC ist ein Oberbegriff, der in der weiteren Gliederung in „Primärelemente" DIN (*primary cells* IEC) und „Sekundärzellen" DIN (*storage cells* IEC; *secondary cells* IEC; *accumulators* IEC) unterteilt wird. *battery* bezeichnet dagegen eine Batterie, die aus mehreren miteinander verbundenen Primärelementen oder Sekundärzellen besteht. In diesem Falle wird sie auch mit *storage battery* („Akkumulator" DIN) bezeichnet.

Dieselbe Unterscheidung gilt für die Begriffspaare *solar cell – solar battery*, *fuel cell – fuel battery* u. a. Dennoch werden in der englischen Fachliteratur die Begriffe *cell* und *battery* bisweilen verwechselt. Oft findet man auch Mischformen wie etwa *solar cell battery*.

Die Benennung *fuel cell* kann entweder eine Brennstoffzelle oder auch ein Brennstoffelement bezeichnen. Im ersten Fall handelt es sich um einen Stromerzeuger zur direkten Energieumsetzung, im zweiten um ein Element eines heterogenen Kernreaktors.

convertor – inverter

convertor /GB/, BS oder *converter* /USA/, ANSI (Umformer IEC, DIN, VDE) ist eine Maschine, die eine Stromart in eine andere umformt, z. B. den Wechselstrom in Gleichstrom oder umgekehrt, oder die die Frequenz (*frequency convertor*) oder Phasenzahl (*phase convertor*) in eine andere ändert.

Oft wird aber – besonders in den USA – das Wort *converter* im engeren Sinne benutzt, und zwar nur für eine umlaufende Maschine, die den Wechselstrom gleichrichtet, um sie vom sog. *inverter* ANSI „Wechselrichter" DIN, VDE zu unterscheiden, der bekanntlich aus Gleichstrom Wechselstrom erzeugt. Manchmal findet man auch die Benennung *inverted rotary convertor* für einen umlaufenden Wechselrichter; *rotary convertor* = „Einankerumformer".

frequency converter ANSI oder *frequency convertor* IEC, auch *frequency changer* BS ist – wie oben angedeutet – eine umlaufende Maschine, die eine Frequenz in *eine* andere umformt. *frequency changer* ist laut ANSI dagegen breiter begriffen und bezeichnet einen Frequenzumformer, der eine Frequenz in *eine oder mehrere* andere Frequenzen umformt.

Der sog. Kaskadenumformer wird in der angelsächsischen Literatur meistens *motor convertor* IEC, BS bzw. *motor converter* ANSI genannt.

Für weitere Bedeutungen von *converter* (in der Funktechnik) siehe auch im Kap. 21 und 25.

energy – power

Während *energy* fast ausnahmslos mit „Energie" oder „Arbeit" (gemessen in J, erg, kcal, kWh, eV) übersetzt werden kann, hat *power* mehrere Bedeutungen: vor allem kann damit auch die Energie bezeichnet werden, wie z. B. *power industry* („Energieerzeugung"). Sonst hat *power* immer die Dimension von W und bezeichnet die Leistung. Oft wird *power* auch mit „Kraft" übersetzt (z. B. *power equipment* „Kraftanlage"), obwohl es sich hier um keine Kraft im physikalischen Sinne (= *force*) handelt.

In Wortverbindungen nimmt *power* folgende Bedeutungen an: „Starkstrom" (*power engineering* „Starkstromtechnik"), „Leistungs-", „End-" (*power stage* „Leistungsstufe", „Endstufe"), „Netz-" (*power pack* „Netzteil" – allerdings nur in Fällen netzgespeister Geräte).

Weitere Bedeutungen, die von Interesse sein mögen: „Potenz" (*power series* „Potenzreihe"), „Vermögen" (*resolving power* „Auflösungsvermögen"), „angetrieben" (*power tool* „Kraftwerkzeug").

input – output – throughput

Im Englischen wird *input* nicht nur im Sinne *input power* (Eingangsleistung), *input voltage* (Eingangsspannung), *input signal* (Eingangssignal) usw. benutzt; vielmehr werden damit Eingangsteile oder Eingangsstufen, namentlich aber die Eingangsklemmen, bezeichnet.

Dasselbe gilt analog für *output*. Unter *output* wird aber oft auch der Ausstoß, die Jahresproduktion *yearly output* (etwa eines Kraftwerkes) verstanden.

Auf dem Gebiete der Informationsverarbeitung wird *input* und *output* „Eingabe" bzw. „Ausgabe" übersetzt (S. Kap. 16).

Im Zusammenhang mit den Wörtern *input* und *output* soll noch auf das Wort *throughput* hingewiesen werden. *throughput* entspricht den deutschen Ausdrücken „Durchgangsleistung" oder „Durchsatz" (auf dem Gebiete der Kernphysik).

7. Transmission and Distribution of Electric Power (Übertragung und Verteilung elektrischer Energie)

7.1. Electric Supply Systems
(Stromversorgungssysteme)

Systems used for distribution of electric power are of varying complexity and consist of (1) a primary *distribution system*, i.e. a network of *transmission lines* carrying three-phase a.c. from the *switch bay* of the plant to a number of *substations*, and (2) a secondary distribution network (Fig. 7–1).

The generating stations are linked to the transmission lines by means of *step-up transformer stations*. The transmission lines are followed by *switching stations* and step-down transformer stations, then *service transformer banks* and, finally, *secondary lines*.

In recent years, the use of high-voltage d.c. transmission has been introduced, with *mercury-arc tubes* as the *converting elements*, but thyristors may be an alternative. In such systems, the power is generated as a.c., stepped up by transformers, *rectified*, transmitted as d.c., *inverted back* to a.c. at the *receiving end* of the line by *inverters*, and stepped down to the *distribution voltage*.

converting element	(hier:) Umformer *m*
distribution system	Verteilungsnetz *n*
distribution voltage	Verteilungsspannung *f*
to invert back	wechselrichten

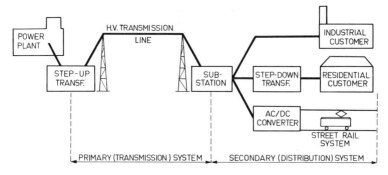

Fig. 7–1. A simplified power transmission and distribution system.
(Vereinfachte Darstellung einer Übertragungs- und Verteilungsanlage.)

AC/DC converter	*Bahnrichteranlage*
H.V. transmission line	*Hochspannungsleitung*
industrial customer	*Industrieverbraucher*
power plant	*Kraftwerk*
primary (transmission) system	*primäres Netz*
residential customer	*Haushaltverbraucher*
secondary (distribution) system	*sekundäres Netz*
step-down transf(ormer)	*Abwärtstransformator*
step-up transf(ormer)	*Aufwärtstransformator*
street rail system	*Straßenbahnsystem*
substation	*Unterstation*
inverter	Wechselrichter *m* VDE
mercury-arc tube	Quecksilberdampfventil *n*
receiving end	Empfangsseite *f*
to rectify	gleichrichten
secondary line	Sekundärleitung *f*
service transformer bank	Gruppentransformatoren *mpl*
step-up transformer station	Aufwärts-Umspannwerk *n*
substation IEC	Unterwerk *n* IEC
switch bay IEC	Schaltfeld *n* IEC
switching station	Verteilerwerk *n*
transmission line	Übertragungsleitung *f*

A selection of related terms — Ausgewählte verwandte Benennungen

control room IEC	Schaltwarte *f* IEC; Kommandoraum *m* /A/, /CH/, IEC
distribution substation IEC	Verteilstation *f* IEC; Netzstation *f* IEC; Verteileranlage *f* IEC
distribution system	(Verteiler)netz *n*
shunt ~ ; parallel ~	Konstantspannung~ IEC; Parallelschaltsystem *n* IEC
series ~ IEC	Konstantstrom~ IEC; Reihen-Schaltsystem *n* IEC; Serienschaltsystem *n* IEC
resonant earthed ~ IEC	gelöschtes ~ IEC; kompensiertes ~ IEC

grid /GB/	britisches Übertragungs- und Verteilungsnetz *n* von 132 kV
holding* IEEE	(hier:) Einhalten *n* der Frequenz
interconnection IEC	Netzkopplung *f* IEC; Netzverbund *m* IEC
network IEC	Netz *n* IEC
meshed ~ IEC	Maschen~ IEC
radial ~ IEC	Strahlen~ IEC
ringed ~ IEC	Ring~ IEC
neutral conductor IEC	Sternpunkt-Leiter *m* VDE, IEC; Nulleiter *m* VDE, IEC; Mittelleiter *m* VDE, IEC
super-grid /GB/	britisches Übertragungsnetz *n* von 275 kV

7.2. Transmission and Distribution Lines Equipment
(Ausrüstung der Übertragungs- und Verteilungsnetze)

In transmission and distribution networks, *overhead lines* are used instead of *underground cables* wherever possible, with conductors *stranded* for flexibility and supported by *pin-type insulators*. For higher voltages, *suspension insulators* are used. The poles and towers should be sufficiently high so that, at the middle of the *span* between them, the bottom of the *wire sag* has sufficient *clearance* over the ground.

Cables have the advantage of greater immunity from *breakdown* and from the effects of weather and lightning. In *L.V. single-phase circuits* common *concentric cables* are often employed, while *oil-filled* or *gas-pressure cables* are used for *H.V. lines*.

An important element of transmission networks is the *high-power transformer* of either *core* or *shell type*, employed in various *star* and *mesh* (*delta* and *wye*) *connections*. Most transformers are *oil-immersed*, and *forced cooling* is sometimes adopted.

For preventing the occurence of faults and *clearing* them, various *protective equipment* is used such as *fuses*, *over-current* and *over-voltage relays*, and *circuit breakers*. *Automatic reclosing* of breakers is frequently used to restore service quickly after a line *trips out* owing to a *transitory fault*.

Lightning arresters are special protective devices designed to discharge electric *surges* resulting from *lightning strokes* or other disturbances which would otherwise *flash over* insulators or *puncture* insulation, resulting in a *line outage* and possible failure of equipment. *Valve-type arresters* which employ a resistance element, and *expulsion-type arresters* with an *arc extinguishing chamber*, are the two general types of these devices.

arc extinguishing chamber	Bogenlöschkammer *f*
automatic reclosing	automatische Wiedereinschaltung *f*
breakdown	(hier:) Aussetzen *n*
circuit breaker	Leistungsschalter *m*
clearance	lichte Entfernung *f*; Abstand *m* VDE
to cleare	(hier:) beseitigen
concentric cable	konzentrisches Kabel *n*
core-type transformer	Kerntransformator *m* VDE
delta connection	Dreieckschaltung *f* DIN, VDE
expulsion-type arrester	Löschrohrableiter *m*
to flash over	überspringen
forced cooling	erzwungene Luftkühlung *f* VDE
fuse	Sicherung *f*

gas-pressure cable	Gaskabel *n*
high-power transformer	Leistungstransformator *m* VDE
H. V. line	Hochspannungsleitung *f*
lightning arrester	Blitzschutz *m*
lightning stroke	Blitzschlag *m*
line outage	Leitungsausfall *m*
L.V. single-phase circuit	Niederspannungs-Einphasennetz *n*
mesh connection	Maschenschaltung *f*
oil-filled cable	Ölkabel *n*
oil-immersed	unter Öl
over-current relay	Überstromrelais *n* DIN, VDE
overhead line	Freileitung *f* VDE
over-voltage relay	Überspannungsrelais *n* DIN, VDE
pin-type insulator	Stützenisolator *m*
protective equipment	Schutzanlage *f*
to puncture	durchschlagen
shell type transformer	Manteltransformator *m* VDE
span	Spannweite *f* VDE
star connection	Sternschaltung *f* DIN, VDE
to strand	verseilen
surge	Stoßspannung *f* DIN, VDE
suspension insulator	Hängeisolator *m*
transitory fault	vorübergehender Fehler IEC
to trip out	ausschalten
underground cable	Erdkabel *n* VDE
valve-type arrester	Ventilableiter *m* VDE
wire sag	Drahtdurchhang *m*
wye connection*	Sternschaltung *f* DIN, VDE

A selection of related terms — **Ausgewählte verwandte Benennungen**

cable duct; cable pipe; cable tube	Kabelschutzrohr *n*
cable joint ANSI; cable splice ANSI	Kabelspleißstelle *f*
cable sheath ANSI	Kabelmantel *m* IEC
composite conductor ANSI	Doppelmetalleiter *m*
conduit ANSI	Kabelleitung *f*
earth IEC, BS; ground ANSI	Erdung *f* VDE
earth current IEC, BS; ground current ANSI	Erdstrom *m*
earth electrode IEE, BS; ground electrode ANSI	Erder *m* VDE
earthing system IEC, BS; grounding system ANSI	Erdungsanlage *f* VDE
insulator	Isolator *m* VDE
~ string	~ kette *f*
lead-in ~	Durchführungs~
strain ~ ; tension ~	Abspann~
lay ANSI	Drallänge *f*
link box	Verteilkasten *m*
multiple conductor	Bündelleiter *m*
switching panel; patch* panel	Schalttafel *f*
troughing	Kabelkanal *m*

twin cable ANSI; double conductor cable IEC	Zweierbündelkabel *n*; paarverseiltes Kabel *n*
voltage	Spannung *f* VDE
contact ~	Berührungs~ VDE
pace ~	Schritt~ VDE

Disturbances and Faults — **Störungen und Fehler**

earth fault IEC	Erdschluß *m* VDE, IEC; Körperschluß *m* IEC
fault IEC	Fehler *m* IEC; Schaden *m* /A/
intermittent ~ IEC	aussetzender ~ IEC; intermittierender ~ IEC
permanent ~ IEC	bleibender ~ ; beständiger ~ IEC
transient ~ IEC	vorübergehender ~ IEC; selbstlöschender ~ IEC
impulse wave IEC	Stoßwelle *f* IEC
chopped ~ ~ IEC	abgeschnittene ~ IEC
full ~ ~ IEC	volle ~ IEC
over-voltage	Überspannung *f* VDE

7.3. Electrical Wiring
(Elektrische Installation)

In *household wiring installations* the conductors connect to a *distribution panel* such lamps and *plug-in sockets* as may be supplied from the *power mains*. A flexible, *steel-armoured cable*, called the BX cable, is used for household wiring more than any other type. Sometimes a cable consisting of two insulated wires in *braided fabric tubing* is used. The cable is usually *drawn in* a rigid metal or flexible plastic *conduit*.

Electrical appliances are connected in parallel across the line consisting of a grounded *neutral wire* and a "*live*" or "*hot*" *lead*. As the line enters the room, it divides into two parallel circuits. One circuit connects a number of *base outlets* (*receptacles*) while the other circuit, with a switch inserted in series, is connected to *lamp sockets*.

In a *factory installation* the bulk of the *power load* is three-phase, while the *lighting circuits* are single-phase. The installation is divided into a number of sections, each of which serves as a *supply centre* for that section and is itself fed from the *main distribution board*. The *mains supply* is brought through a 4-core cable from the nearest transformer station, and connected to the *bus-bars* of the main distribution board.

base outlet /USA/, /GB/; receptacle /USA/	Steckdose *f* DIN, VDE
braided fabric tubing	Rohr *n* aus geflochtenem Gewebe
bus-bar	Sammelschiene *f* VDE
conduit CEE	Installationsrohr *n* VDE
distribution panel	Verteilungstafel *f*
to draw in	einziehen
electrical appliances	elektrische Haushaltgeräte *npl*
factory installation	(hier:) Industrieinstallation *f*
household wiring installation	elektrische Hausinstallation *f* VDE
lamp socket; lamp holder	Lampenfassung *f* DIN, VDE
lighting circuit	Lichtstromkreis *m*
"live" lead; "hot" lead	unter Spannung stehender Leiter *m* VDE
main distribution board	Verteilungstafel *f*
mains supply	(hier:) Netzzuführung *f*

neutral wire; neutral conductor CEE	Mittelleiter *m* DIN, VDE
plug-in socket	Steckdose *f* DIN, VDE
power load	Last *f*
power mains	Netz *n*
steel-armoured cable	Stahlpanzerkabel *n*
supply centre	Quelle *f*

A selection of related terms	**Ausgewählte verwandte Benennungen**
conduit box	Abzweigkasten *m*
plug	Stecker *m*
branch ~	Abzweig~
two-pole protection pin ~	~ mit Schutzkontakt
plug adapter	Übergangsstecker *m*
plug and socket	Steckvorrichtung *f* VDE
wiring installation	(Leitungs)verlegung *f*
buried ~ ; concealed ~ ; flush(ed) ~ *	Unterputz~
surface ~ *	Aufputz~

Fachsprachliche Anmerkungen

flush – surface (mounting)

Für versenkte, bündig abschneidende, eingelassene, Unterputz-Geräte werden ebenso viele englische Benennungen gebraucht: *flush, flushed, panel, recessed, sunk, buried* (*switches, instruments, etc.*). Die britische Norm BS 205 betrachtet die meisten von ihnen als gleichwertig, in den Katalogen der Hersteller werden aber zwischen ihnen Grenzen gezogen.

Für die Außen- oder Aufputzmontage gibt es dagegen im Englischen nur das Wort *surface mounting*.

patch

Das Verb *to patch* heißt einmal „stöpseln, Verbindungen in einer Schalttafel einstöpseln". Von hier leiten sich folgende Wortverbindungen her:

patch bay	Verbindungsfeld
patchboard	Stecktafel, Buchsenfeld
patch cord	steckbare Verbindungsleitung
patch panel	Schalttafel
patching jackfield	Klinkenfeld, Stöpselfeld

Zum anderen bedeutet *to patch* soviel wie *to erase* „überlochen" GAMM (unwirksam machen einer Information auf Lochstreifen oder Lochkarte), aber auch korrigieren (z. B. Programm im Kernspeicher).

pull-in – lock-in – hold – sync

pull-in ANSI, auch *pulling-in* oder einfach *pulling* bezeichnet ganz allgemein die gewollte oder ungewollte Änderung der Frequenz eines Senders, Generators oder Oszillators als Folge z. B. einer Laständerung oder durch Mitnahme oder Mitziehen der Frequenz des Generators durch eine fremde Frequenz. Diese Benennung hat jedoch noch mehrere andere Bedeutungen (beim Relais: „Ansprechen"; bei Elektronenröhren: „Lastverstimmung").

lock-in ist beinahe synonym mit *pull-in* und bezeichnet die gewollte Synchronisierung einer Frequenz oder Schwingung, die vom Sollwert abweicht, durch ein Bezugssignal (z. B. Horizontal-Synchronimpulse im Fernsehen). Mit *lock-in* zusammengesetzte Wortverbindungen werden im solchen Spezialfall mit „Fang-" übersetzt (z. B *lock-in range* „Fangbereich"). Alle elektrotechnischen Benennungen und Ausdrücke die den Stamm *lock* enthalten sind übrigens von dem aus der Mechanik bekannter Begriff „Sperre, Einrasten, Verriegelung, Blockierung" abgeleitet. So bedeutet z. B in der Radartechnik *locking* das „automatische Nachlaufen eines Zieles", d. h. Verknüpfung der Bewegung der Radaranlage mit der des Zieles.

hold bezeichnet den Zustand, in dem eine synchronisierte Größe synchron bleibt: *hold range* „Haltebereich". In der Starkstromtechnik bedeutet *holding* IEEE den Zustand, wo ein Generator oder Kraftwerk die Frequenz auch bei Laständerungen einhält. Oft findet man aber das Wort *hold* in einer Beziehung, wo laut Definition eher *lock-in* angemessen wäre, z. B. *field hold control* „Bildfang" (im Fernsehen).

synchronization (abgekürzt *sync*) kann als Oberbegriff für *locking-in* und *hold* angesehen werden, bezeichnet aber das Verfahren („Synchronisierung"), nicht den Zustand (= *synchronism*).

Die Kurzbenennung *sync* wird oft für *synchronizing, synchronized* oder *synchronization*, nie für andere Benennungen desselben Stammes (etwa wie *synchronous = synchro*!) benutzt: *sync level* „Synchronwert", aber *synchronous detector* „Synchrondetektor".

wye connection – Y connection

Der Begriff „Sternschaltung" könnte mit Hilfe des Schaltkurzzeichens Y (für Stern) im Deutschen auch als „Y-Schaltung" wiedergegeben werden, was aber nicht geschieht. Dagegen hat sich im Englischen neben dem Wort *star connection* IEC das Wort *Y-connection* /USA/ eingebürgert. Darüber hinaus hat die englische Benennung des Buchstaben Y [wai] dazu geführt, daß besonders im Amerikanischen das Zeichen Y durch das Wort *wye* (lies [wai]) ersetzt worden ist, z. B. *wye connection* (Sternschaltung), *wye-grounded* (in Sternschaltung geerdet).

8. Electrical Machinery (Elektrische Maschinen)

8.1. Parts of Electrical Machines (Teile elektrischer Maschinen)

One of the basic elements of *electrical machines* is the *winding*, consisting of a number of *turns* and performing various functions in the operation (the *armature winding* in which the principal conversion of energy occurs, as well as the *excitation winding*, the *commutating winding*, etc.).

Some of the main forms of winding construction are the concentrated winding, i.e. that of a field system with *salient poles* or with one *slot* per *pole*, the *distributed winding*, occupying several slots per pole, the ring winding, and the *drum winding*. The type termed *squirrel-cage winding* consists of a number of conducting bars having their extremities connected by metal rings or plates at each end.

Other fundamental elements of electrical machinery are the fixed and rotating parts termed a *stator* and a *rotor*, respectively, various types of poles (for example the *field pole* placed between the *yoke* and the *air gap*, and its variety, the salient pole projecting from the yoke or *hub* towards the *armature*). Some of the other main elements are the *pole shoes* shaping the air gap, *magnetic cores*, slots, *spiders*, and *brushes*. The stationary parts of an electric machine are supported by a *frame*.

air-gap IEC	Luftspalt *m* DIN, IEC
armature IEC	Anker *m* IEC
armature winding IEC	Ankerwicklung *f* IEC
brush IEC	Bürste *f* IEC
commutating winding IEC	Wendefeldwicklung *f* IEC
distributed winding IEC	verteilte Wicklung *f* IEC
drum winding IEC	Trommelwicklung *f* IEC
electrical machine IEC	elektrische Maschine *f* VDE, IEC
excitation winding IEC	Erregerwicklung *f* IEC
field pole IEC	Feldpol *m* IEC
frame IEC	Gehäuse *n* IEC
hub	Nabe *f*
magnetic core IEC	Magnetkern *m* IEC
pole IEC	Pol *m* IEC
pole shoe IEC	Polschuh *m* IEC
rotor IEC	Läufer *m* VDE, IEC; Rotor *m* IEC
salient pole IEC	ausgeprägter Pol *m* IEC
slot IEC	Nut *f* IEC
spider IEC	Läuferstern *m* IEC; Rotorstern *m* IEC
squirrel-cage winding IEC	Käfigwicklung *f* IEC
stator IEC	Ständer *m* VDE, IEC; Stator *m* IEC
turn* IEC	Windung *f* IEC
winding* IEC	Wicklung *f* VDE, IEC
yoke IEC	Joch *n* IEC

A selection of related terms / Ausgewählte verwandte Benennungen

commutator IEC	Stromwender *m* IEC; Kommutator *m* IEC
commutator lug /GB/, IEC; commutator riser /USA/, IEC	Stromwenderfahne *f* IEC
commutating pole IEC; interpole IEC	Wendepol *m* IEC
phase spread /GB/, IEC; phasebelt /USA/, IEC	Zonenbreite *f*
pitch IEC	Schritt *m* IEC
pitch factor IEC	Sehnungsfaktor *m* IEC
pole face IEC	Pol(schuh)fläche *f* IEC
pole horn IEC	Polkante *f*
slip ring IEC	Schleifring *m* IEC
tier IEC	(Wicklungs)etage *f* IEC; Ebene *f* IEC
winding* IEC	Wicklung *f* IEC
control ~ IEC	Steuer~ IEC
diamond ~ IEC	Gleichspulen~ IEC
doughnut*~	Torus~
end ~ IEC	Spulenkopf *m* IEC; Stirnverbindung *f* IEC
fractional-pitch ~ IEC	Sehnen~ IEC
fractional-slot ~ IEC	Bruchloch~ IEC
full-pitch ~ IEC	Durchmesser~ IEC
integer-slot ~ IEC	Ganzloch~ IEC
lap ~ IEC	Schleifen~ IEC
one-position ~ IEC; single-layer ~ IEC	Einschicht~ IEC; einlagige ~ IEC

stepped ~ IEC; split ~ IEC Treppen~ IEC
wave ~ IEC Wellen~ IEC

8.2. Rotating Electric Machines
(Umlaufende elektrische Maschinen)

The most used *rotating electric machine*, the electric motor, is a machine which, receiving electric energy, *converts* it into mechanical energy by producing a *turning effort*, or *torque*. A classification of the principal types of motors given in Fig. 8–1 shows a primary division into *alternating-current* and *direct-current motors*.

The *synchronous motor* is practically an *alternator operated inverted*. If it is loaded to where it *lags behind* the synchronous speed, it quickly *falls out of step* and comes to rest. In an *induction motor* a primary winding on one member (usually the stator) is connected to the power source, and a *polyphase* secondary winding or a *squirrel-cage* secondary winding on the other member (usually the rotor) carries induced current.

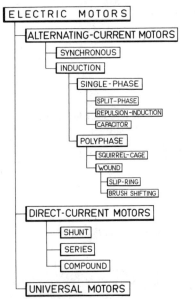

Fig. 8–1. A Classification of electric motors.
(Einteilung elektrischer Motoren)

alternating-current motor	Wechselstrommotor
brush shifting motor	Verstellmotor
capacitor motor	Kondensatormotor
compound motor	Verbundmotor
direct-current motor	Gleichstrommotor
induction motor	Induktionsmotor
polyphase motor	Mehrphasen-Motor
repulsion-induction motor	Repulsions-Induktionsmotor
series motor	Nebenschlußmotor
shunt motor	Reihenschlußmotor
single-phase motor	Einphasenmotor
slip-ring motor	Schleifringmotor
split-phase motor	Motor mit Spaltphase, Spaltmotor
squirrel-cage motor	Käfig(anker)motor
synchronous motor	Synchronmotor
universal motor	Universalmotor
wound motor	Asynchronmotor mit gewickeltem Anker

Most a.c. industrial motors are three-phase, but a considerable number of single-phase *fractional horse-power* (f.h.p.) motors are used for simplification of *wiring systems*.

A *universal motor* is a *series motor* which may be operated on either a.c. or d.c. It is a *series-wound* or *compensated* series-wound motor designed to operate at approximately the same speed and *output* on either d.c. or single-phase a.c. of approximately the same *r.m.s. voltage*.

alternating-current motor IEC Wechselstrommotor *m* IEC
alternator IEC Synchrongenerator *m* IEC

compensated	kompensiert
to convert	umformen; (hier:) umwandeln; umsetzen
direct-current motor IEC	Gleichstrommotor m IEC
to fall out of step	außer Tritt fallen
fractional horse-power motor*; f.h.p. motor*	Kleinmotor m
induction motor* IEC	Induktionsmotor m IEC
to lag	zurückbleiben; nacheilen
operated inverted	(hier:) umgekehrt arbeitend; umgekehrt betrieben
output	(hier:) Leistung f
polyphase	Mehrphasen-
r.m.s. voltage	effektive Spannung f
rotating electric machine	umlaufende elektrische Maschine f
series motor	Reihenschlußmotor m
series-wound motor	reihenschlußgewickelter Motor m
squirrel-cage	Käfig m
synchronous* motor IEC	Synchronmotor m IEC
torque	Drehmoment n
turning effort	Drehmoment n
universal motor	Allstrommotor m
wiring system	(hier:) elektrische Installation f

A selection of related terms

Ausgewählte verwandte Benennungen

convertor IEC, /GB/; converter ANSI	Umformer m IEC
excitation IEC	Erregung f IEC
compound ~ IEC	Doppelschluß~ IEC; Verbund~ IEC
cumulative compound ~ IEC	Mitverbund~ IEC
level-compound ~ IEC	Verbund~ für gleichbleibende Spannung IEC; Flachverbund~ IEC
over-compound ~ IEC	Überverbund~ IEC
separate ~ IEC	Fremd~ IEC
motor IEC	Motor m IEC
doubly-fed polyphase shunt commutator ~ IEC	ständergespeister Mehrphasen-Nebenschluß~ IEC
integral ~ IEC	Groß~ IEC
polyphase compound commutator ~ IEC	Mehrphasen-Reihenschluß~ mit Begrenzung der Leerlaufdrehzahl IEC
polyphase series commutator ~ with rotor transformer IEC	Mehrphasen-Reihenschluß~ mit Zwischentransformator IEC
polyphase shunt commutator ~ IEC	Mehrphasen-Nebenschluß~ IEC
reciprocating (movement) ~ IEC	Schwing~ IEC
self-compensated ~ (with primary rotor) IEC	läufergespeister kompensierter Induktions~ IEC
shunt-characteristic polyphase commutator ~ with double set of brushes IEC; Schrage ~ IEC	läufergespeister Mehrphasen-Nebenschluß-Kommutator~ IEC
single-phase commutator ~ with self-excitation IEC	kompensierter Repulsions~ IEC
single-phase commutator ~ with series compensating winding IEC	Einphasen-Reihenschluß~ mit Kompensationswicklung IEC

single-phase series commutator ~ with short-circuited compensating winding IEC	Einphasen-Reihenschluß~ mit kurzgeschlossener Kompensationswicklung IEC
step ~ IEC	Schritt~ IEC
stalled motor*	durch Überlasten blockierter Motor *m*
synchro*	Drehmelder *m* DIN

8.3. Electric Traction
(Elektrotraktion, hier: Elektrische Zugförderung)

The application of electricity to the operation of transport vehicles covers a very wide field, ranging from heavy *mainline railway locomotives* to small *battery-operated trucks* for factory and warehouse transport. *Traction* falls into three broad divisions, viz. railways, road transport, and *industrial service*.

For *d.c. traction* service it is standard practice to use series-wound motors on account of their favourable *torque/speed characteristic* and their stability against variations in supply voltage. They have also a high *overload-torque capacity*.

Besides d.c. and *battery traction*, three a.c. systems are in use on railways:

(1) Three-phase, $16^2/_3$ Hz systems with two *overhead wires* and *rail return*. Speed control is obtained by *pole-changing connection* of the motors.

(2) Single-phase, $16^2/_3$ or 25 Hz systems with one overhead wire and rail return. Line voltages up to 22,000 V are employed with step-down transformers at locomotives and *motor coaches*.

(3) Single phase, 50 Hz systems with three-phase locomotive motors fed by a phase convertor, or using *rectifier locomotives*, or employing series-type 50 Hz motors.

battery-operated truck	Akku(mulator)wagen *m*
battery traction; battery electric traction IEC	Akkumulatoren-Zugförderung *f* IEC
d.c. traction	Gleichstrom-Zugförderung *f*
industrial service	(hier:) Werkverkehr *m*
mainline	Hauptstrecke *f*
motor coach IEC; motor car	Triebwagen *m* IEC
overhead wire	Freileitung *f* VDE
overload-torque capacity	(hier:) Überlastbarkeit *f*
pole-changing connection	Polumschaltung *f*
rail return	Schienenrückleitung *f*
railway locomotive	Eisenbahnlokomotive *f*
rectifier locomotive	Stromrichterlokomotive *f*
torque/speed characteristic	Drehmomentkennlinie *f*
traction	Zugförderung *f*; (Be)Förderung *f*
A selection of related terms	**Ausgewählte verwandte Benennungen**
catenary line	Kettenfahrleitung *f*
conductor rail IEC	Stromschiene *f* IEC
contact wire IEC	Fahrdraht *m* IEC
current collector IEC	Stromabnehmer *m* IEC
drive IEC	Antrieb *m* VDE, IEC
effort IEC	Zugkraft *f* IEC
braking ~ IEC	Bremskraft am Radumfang IEC
holding braking ~ IEC	Bremskraft bei Gefällebremsung IEC
tractive ~ IEC	~ am Radumfang IEC
field weakening IEC	Feldschwächung *f* IEC

~	tilde
√	root sign; radical sign
|	modulus bars
( )	round brackets; parentheses
(	bracket open
)	bracket close
[ ]	square brackets
{ }	braces
#	number; No; item
∞	infinity sign
≃	congruence sign

3. Sprechweise mathematischer Ausdrücke

1–9	one to nine; one through nine
$1/2$	a (one) half
$3/4$	three quarters; three fourths
$25/58$	twenty-five fifty-eights; twenty five over fifty eight
$10\frac{1}{5}$	ten and one fifth
0.2	o [ou] point two; zero point two; nought point two
.5	point five
1.0026	one point zero zero two six
$a = b$	a equals b; a is equal to b ISO, UIP
$a + b$	a plus b ISO, UIP
$a - b$	a minus b ISO, UIP
$a \pm b$	a plus or minus b UIP
$ab; a \times b$	a times b; a multiplied by b ISO, UIP
$a \div b$	a divided by b ISO, UIP
$\frac{a}{b}$	a over b; a divided by b ISO, UIP
$\frac{ab}{c+d}$	a times b over c plus d
a/b	a solidus b; a slant b
$a : b$	a is to b
$a : b :: c : d$	a is to b as c is to d
$a \propto b$	a is proportional to b ISO; Vorzugszeichen $a \sim b$ ISO
$a \sim b$	a is asymptotically equal to b ISO; Vorzugszeichen $a \cong b$ ISO
$a \approx b$	a is nearly equal to b, a is approximately equal to b
$a \equiv b$	a is identically equal to b ISO
$a \not\equiv b$	a is not identical with b
$\therefore a = b$	therefore a equals b
$\because a = b$	because a equals b
$a \to b$	a tends to b; a approaches b ISO, UIP
$a!$	factorial a ISO; a factorial
$a < b$	a is less than b; a is smaller than b ISO
$a > b$	a is greater than b; a is larger than b ISO
$a \ll b$	a is much less than b; a is much smaller than b ISO
$a \gg b$	a is much greater than b; a is much larger than b ISO
$a \geqslant b$	a is larger than or equal to b ISO
a_b	a subscript b; a sub b
a'	a dash; a prime
a''	a double dash; a double prime; a second prime
a^2	a squared

a^3	a cubed; a to the third power
a^n	a raised to the power n ISO
a^{-n}	a to the minus nth power; a to minus n
a^{bx}	a to the power bx
$a^{\frac{1}{4}}$	a to the power one-fourth
$\sqrt{a}$	square root of a
$\sqrt[3]{a}$	cube root of a
$\sqrt[n]{a}$	nth root of a ISO
$\sqrt{a^2 + b^2}$	square root out of a squared plus b squared
$\|a\|$	modulus of a; absolute value of a; magnitude of a ISO
$\begin{vmatrix} a_{11} & a_{12} \\ a_{21} & a_{22} \end{vmatrix}$	determinant (matrix): *first row:* a sub one one, a sub one two; *second row:* a sub two one, a sub two two
$\binom{n}{k}$	binomial n over k
$\sum_{a=1}^{n}$	the sum from a equals one to n
$\prod_{a=1}^{n}$	the product from a equals one to n
$\sin \alpha$	sine α; sine of α ISO
$\cos \alpha$	cosine α; cosine of α ISO
$\tan \alpha$	tan α (Ausspr.: [ten alpha]); tangent α; tangent of α ISO
$\sec \alpha$	secant α; secant of α ISO
$\sinh \alpha$	shine α (Ausspr.: [schein alpha]); hyperbolic sine of α ISO
$\cosh \alpha$	cosh α (Ausspr.: [kosch alpha]); hyperbolic cosine of α ISO
$\text{sech } \alpha$	sech α (Ausspr.: [setsch alpha]); hyperbolic secant of α ISO
$\cos^{-1} \alpha$	inverse cosine α; cos minus one α
$\log x$	log of x
$\log_a x$	logarithm to the base a of x ISO
d	differentiation sign
dx/dy	dx by dy (derivative of x with respect to y)
$\dot{x}$	first derivative of x ⎫
	⎬ with respect to time
$\ddot{x}$	second derivative of x ⎭
δx	variation of x; delta x ISO
$\partial x/\partial a$	partial derivative of x with respect to a
$\lim f(x)$	limit of the function of x
$\int_{x=a}^{b} f(x) dx$	integral between the limits a and b of x dx; definite integral of f(x) from x = a to x = b
$\iint$	double integral
$\iiint$	triple integral
$\oint$	circuital integral; integral round a closed circuit
$\int f(x)dx$	indefinite integral of f(x)dx

Beispiele:

$\dfrac{a'}{b} c_d - e$	a prime over b times c sub d minus e; a prime over b, this fraction multiplied by c sub d minus e.

$\int \dfrac{dy}{\sqrt{c^2 - y^2}}$ (indefinite) integral of dy over the square root out of c square minus y square.

$v = u\sqrt{\sin^2 i - \cos^2 i}$ v is equal to u times square root of sine square i minus cosine square i.

$\therefore M_t = G\vartheta \dfrac{\pi d^4}{32}$ therefore M sub t is equal to G theta times πd to the fourth power divided by thirty-two.

4. Boolesche Verknüpfungen (Boolean operations)

Im folgenden werden die englischen und deutschen Benennungen für boolesche Verknüpfungen sowie deren Schreibweisen gegenübergestellt. Da es für jede Verknüpfungsart eine große Anzahl von englischen Benennungen gibt, werden diese in einer ungefähr nach der Verwendungshäufigkeit absteigenden Reihenfolge aufgezählt. Bei den englischen Benennungen wird auch die Sprechweise angegeben.

Verknüpfung	Schreibweise	Definition	Sprechweise	Verknüpfung	Schreibweise nach DIN
Einstellige Boolesche Verknüpfung (Single Operand Boolean Operation)					
1. negation; not operation; (boolean) complementation; inversion	$y = \bar{a}$ $\sim a$ $(-A)$ Na a'	a 0 1 ——— y 1 0	not a	Negation DIN	$y = -a$ $\bar{a}$ Sprechweise: nicht a; a nicht; a negiert
Zweistellige Boolesche Verknüpfungen (Dyadic Boolean Operations; Boolean Connectives)					
2. or-operation BS; logical sum; inclusive-or operation; disjunction; union; join; alteration.	$y = a \vee b$ $a \cup b$ $a + b$ $A\ ab$	a 0101 b 0011 ——— y 0111	a or b	ODER-Verknüpfung DIN; Disjunktion DIN	$y = a \vee b$ Sprechweise: a oder b
3. and-operation BS; logical product; conjunction; intersection; collation; meet.	$y = a\ \&\ b$ $a \wedge b$ $a \cap b$ $a \cdot b$ ab $K\ ab$	a 0101 b 0011 ——— y 0001	a and b	UND-Verknüpfung DIN; Konjunktion DIN	$y = a \wedge b$ Sprechweise: a und b

Boolesche Verknüpfungen (Boolean operations)

Verknüpfung	Schreibweise	Definition	Sprechweise	Verknüpfung	Schreibweise nach DIN
4. nor-operation BS; neither-or operation; dagger operation /USA/; rejection; Peirce function; zero match; joint denial.	$y = \overline{(a \vee b)}$	a 0101 b 0011 ――― y 1000	a nor b; neither a nor b	NOR-Verknüpfung DIN	$y = a \overline{\vee} b$
5. nand-operation; not-and operation BS; nonconjunction; not-both operation; Sheffer stroke; Sheffer function; alternative denial operation; dispersion.	$y = \overline{(a \& b)}$ $a \wedge b$ $a \mid b$ D ab	a 0101 b 0011 ――― y 1110	not ab	NAND-Verknüpfung DIN	$y = a \overline{\wedge} b$
6. equivalence operation BS; bi-conditional operation; match.	$y = a \equiv b$ E ab $(a \wedge b) \vee (\bar{a} \wedge \bar{b})$ $a \rightleftharpoons b$	a 0101 b 0011 ――― y 1001	a equivalent to b; a if, and only if, b	Äquivalenz DIN	$y = a \equiv b$
7. non-equivalence operation BS; exclusive or operation; modulo 2 sum; addition without carry; exjunction; distance; symmetric difference; diversity.	$y = a \neq b$ $a \oplus b$ R ab $(a \wedge \bar{b}) \vee (\bar{a} \wedge b)$	a 0101 b 0011 ――― y 0110	a not equivalent to b	Antivalenz DIN	$y = a \neq b$
8. inclusion; conditional implication operation BS; if-then operation; material implication.	$y = \bar{a} \vee b$ $a \leq b$ $a \supset b$ C ab	a 0101 b 0011 ――― y 1011	if a then b	Implikation DIN	$y = a \supset b$

Boolesche Verknüpfungen (Boolean operations)

Verknüpfung	Schreibweise	Definition	Sprechweise	Verknüpfung	Schreibweise nach DIN
9. exclusion; not-if then operation BS; difference; sub-junction; sine-junction.	$y = a \& \bar{b}$ $a(-b)$ $a \wedge \bar{b}$	a 0101 b 0011 ――― y 0100	if a then not b	Inhibition DIN	–

Bemerkung: In den angelsächsischen Ländern ist die Schreibweise 0;1 für dyadische Zahlen üblicher als O;L.

5. Mengenlehre: Fundamentale Symbole und Sprechweise

$a \in S$	a is an element of S
$a \notin S$	a is not an element of S
$S = (a, b)$	set S with elements a, b
$S = \emptyset$	empty set S; S is an empty set
$U \subseteq V$	U is a subset of V; U is contained within V
$U \subset V$	U is a proper subset of V; U is contained properly in V
$U \cup V$	sum of U and V; union of U and V; join of U and V
$U \cap V$	intersection of U and V; product of U and V; meet of U and V
$U \times V$	cross product of U and V; Cartesian product of U and V
$U \sim V$	U is equivalent to V

6. Grundbegriffe der Geometrie

point	Punkt
intersection	Schnittpunkt
vertex	Scheitelpunkt
straight line	Gerade
given straight line	Strecke
axis (of symmetry)	(Symmetrie)achse
parallel	Parallele
perpendicular; normal	Senkrechte; Normale
angle	Winkel
right ~	rechter ~
acute ~	spitzer ~
obtuse ~	stumpfer ~
arm (of the angle)	Schenkel (des Winkels)

Plane Figures and Curves   **Ebene Figuren und Kurven**

triangle	Dreieck
~ side	Seite des -s
~ altitude	Höhe des -s
~ median	Mittellinie des -s
right-angled ~	rechtwinkliges ~

isosceles ~	gleichschenkliges ~
equilateral ~	gleichseitiges ~
curve	Kurve
curvature	Krümmung
radius of ~	Krümmungs-Radius
circle	Kreis
~ radius	Halbmesser des -es
~ diameter	Durchmesser des -es
~ circumference	Umfang des -es
~ segment	Kreisabschnitt
~ sector	Kreisausschnitt
~ arc	Kreisbogen; Bogen
tangent	Tangente
secant	Sekante
chord	Sehne
ring; annulus	Kreisring; Ringfläche
coordinates	Koordinaten
orthogonal ~	kartesische ~
polar ~	Polarkoordinaten
abscissa	Abszisse
ordinate	Ordinate
coordinate axis	Koordinatenachse
locus	geometrischer Ort
conic section	Kegelschnitt
ellipse	Ellipse
parabole	Parabel
hyperbole	Hyperbel
focus	Brennpunkt
asymptote	Asymptote
spiral	Spirale
loop	Schleife
cycloid	Zykloide
catenary	Kettenlinie
quadrilateral	Vierseit
rectangle	Rechteck
square	Quadrat; Viereck
rhombus	Rhombus
rhomboid	Rhomboid
parallelogram	Parallelogramm
kite	Deltoid
trapezium	Trapez
trapezoid	unregelmäßiges Viereck
polygon	Vieleck

Space Curves and Solids / Räumliche Kurven und Körper

helix	Wendel; Schraubenlinie
prism	Prisma
parallelepiped	Parallelepiped
cube	Würfel
pyramid	Pyramide
truncated ~ ; frustum of a ~	Pyramidenstumpf
body of revolution	Drehkörper

cylinder	Zylinder
curved surface of a ~	Mantel des -s
generating line	Mantellinie
base	Stirnfläche
cone	Kegel
truncated ~ ; frustum of a ~	-stumpf
sphere	Kugel
spherical cap (solid and surface)	Kugelabschnitt; Kalotte; Kugelkappe
spherical segment (solid)	Kugelschicht
spherical zone (surface)	Kugelzone
great circle	Großkreis
ellipsoid	Ellipsoid
prolate ~	verlängertes ~
oblate ~	abgeplattetes ~
paraboloid	Paraboloid
hyperboloid	Hyperboloid
one sheet ~	einschaliges ~
two sheet ~	zweischaliges ~

ANHANG:

Über fachsprachliche Wendungen (verfaßt von Dr.-Ing. A. Warner)

Mit dem vorliegenden Buch wird versucht, die englische Fachsprache der Elektrotechnik durch Lesen von Fachtexten zu vermitteln – selbstverständlich unter Zuhilfenahme von beigefügten Wortverzeichnissen und fachsprachlichen Anmerkungen. Man wird nebenbei feststellen, daß in den Fachtexten die facheigenen Begriffe meistens von Substantiven, die gemeinsprachlichen Begriffe meistens von Verben verkörpert werden. Mit anderen Worten, fachliche Texte können in Wortgruppen mit verbalem Oberglied (Verbalgruppen) aufgelöst werden, an denen sich verschiedenartige Unterglieder (z. B. Partizip, Substantiv, Adjektiv) nachweisen lassen. Da aber fachliche Texte auch als Folgen von in grammatischen Sätzen ausgedrückten Gedanken über technisch-wissenschaftliche Sachverhalte aufzufassen sind, können neben den Verbalgruppen auch Sätze zur Aufgliederung herangezogen werden. Es darf jedoch nicht übersehen werden, daß für fachliche Zwecke nicht ausreichend fachsprachliche Verben zur Verfügung stehen und daß daher auf gemeinsprachliche Verben zurückgegriffen werden muß.

Dieser Gedankengang, den der Autor in seiner Dissertation „Internationale Angleichung fachsprachlicher Wendungen der Elektrotechnik; Versuch einer Aufstellung phraseologischer Grundsätze für die Technik" (Berlin: VDE-Verlag 1966) begründet, führt zum Begriff „fachsprachliche Wendung" (kurz: Fachwendung). Unter „fachsprachlicher Wendung" soll eine – getrennt in einzelnen Wörtern geschriebene – Verbindung von Substantiv mit Verb verstanden werden, die in ihrer einfachsten grammatischen Form vorkommt als „Infinitivform des Verbs + Substantiv" (Verbalgruppe) oder als „Substantiv + Personalform eines Verbs" (Satz); Beispiele: *ein Gerät einschalten* bzw. *der Strom fließt*. Ein Wörterbuch, das fachsprachliche Wendungen enthält, heißt „phraseologisches Wörterbuch".

Je nach der Vertrautheit mit der betreffenden Fachsprache greift der Sprecher (oder Verfasser) auf seinen aktiven Fachwortschatz (Gedächtnis), auf das betreffende Fachschrifttum (siehe die Textauswahl dieses Buches) oder auf ein phraseologisches Wörterbuch des jeweiligen Faches zurück.

In einem gewissen Umfang möchte dieses Buch auch ein phraseologisches Wörterbuch sein. Es werden nämlich fachsprachliche Wendungen von fünf Themen dargeboten: *Prüfungen und Versuche; Phasenverschiebung; Abstimmen, Resonanz, Verstimmen; Relais; Ablesung von Meßwerten*. Dabei ist folgendes zu beachten:

Die fachsprachlichen Wendungen derselben laufenden Nummer sind gleichbedeutend. Ebenso sind es die fremdsprachigen Entsprechungen.

Die eckige Klammer bedeutet, daß der eingeschlossene Teil das zwischen dem Grenzzeichen ⌐ und der eckigen Klammer stehende Wort ersetzen darf.

Eine runde Klammer bedeutet, daß der eingeschlossene Teil weggelassen werden darf, wenn der Sinnzusammenhang Unklarheit ausschließt.

Prüfungen und Versuche"

1 das Gerät einer Prüfung ⌈unterziehen [unterwerfen]	to ⌈submit [subject] the apparatus to a test
2 Prüfungen an Mustern ⌈durchführen [anstellen]	to carry out tests on samples; to make tests on samples; to do tests on samples; to execute tests on samples
3 die Muster werden gerade geprüft; die Muster durchlaufen gerade die Prüfung	the samples are ⌈running on test [being tested]
4 mit den Prüfungen beginnen	to start tests
5 die Prüfung ⌈beginnt [fängt an]	the test ⌈commences [starts]
6 die Prüfung ⌈fortsetzen [fortführen]	to continue the test; to maintain the test; to go on testing
7 eine 16tägige Versuchsreihe abschließen	to complete a run of 16 days
8 die Güte überprüfen	to check the quality
9 die Spannung des Gerätes ⌈überprüfen [nachsehen; nachprüfen]	to check the voltage of the apparatus
10 die Prüfung ist hart	the test is ⌈severe [rigorous]
11 das Gerät widersteht der Beanspruchung; das Gerät ⌈verträgt [bewältigt] die Beanspruchung; das Gerät hält die Beanspruchung aus	the apparatus withstands the stress; the apparatus stands up to the stress
12 der Transistor muß Temperaturen bis 150 °C vertragen	the transistor must withstand temperatures up to 150 °C
13 der Transistor kann eine Verlustleistung von 5 W vertragen	the transistor can tolerate 5-watt dissipation
14 die Transistoren werden bei hohen Temperaturen betrieben	the transistors will be operated ⌈at [in] high temperatures

„Phasenverschiebung"

1 die Größen sind gegeneinander phasenverschoben	the quantities show a phase displacement between them
2 eine Größe eilt einer anderen Größe nach	one quantity lags behind another quantity; one quantity lags in phase in relation to another quantity
3 die Spannung ⌈ist [liegt] mit dem Strom in Phase	the voltage is in phase with the current
4 die beiden Spannungen sind ⌈gleichphasig [phasengleich] [in Phase]	both voltages are co-phasal [equiphase] [in phase]
5 eine Größe eilt einer anderen Größe vor	one quantity leads behind another quantity; one quantity leads in phase in relation to another quantity
6 die Spannung eilt dem Strom vor	the voltage leads (behind) the current
7 die Größen sind um ⌈90° [$\pi/2$]phasenverschoben	the quantities have a phase displacement of $\pi/2$ between them
8 die Spannung ist gegenüber dem Strom um 90° (phasen)verschoben	the voltage is in quadrature with the current

„Abstimmen, Resonanz, Verstimmen"

1 einen Schwingkreis abstimmen — to tune a circuit
2 einen Schwingkreis in Resonanz bringen — to tune a circuit for resonance
3 den Meßempfänger um 12 kHz verstimmen — to tune the meter away by 12 kHz
4 der Meßempfänger ist um 12 kHz verstimmt — the meter is 12 kHz ⌈off tune [out of tune]
5 einen UHF-Kanal wählen — to select a UHF channel
6 den Fernsehempfänger auf den UHF-Bereich einstellen — to set the TV set at UHF position
7 den Fernsehempfänger auf Kanal 5 schalten — to tune the TV set to channel 5
8 die Empfänger sind auf einen deutschen Sender abgestimmt — the receivers are tuned to a German station
9 Sie hören den AFN — you are tuned to AFN
10 der Schwingkreis ist auf die Frequenz f des Oszillators abgestimmt — the circuit is tuned to the frequency f of the oscillator
11 der Schwingkreis ist bei der Frequenz f in Resonanz — the circuit vibrates by resonance with the frequency f
12 der Schwingkreis ist mit einem anderen Schwingkreis in Resonanz — the circuit vibrates by resonance with another circuit

„Relais"

1 das Relais ⌈erregen [betätigen; zum Ansprechen bringen; zum Anzug bringen; bringen] — to actuate the relais; to trip the relay; to activate the relay; to energize the relay; to operate the relay
2 das Relais ⌈ spricht an [zieht an; wird erregt] — the relay ⌈ pulls up [picks up]
3 das Relais schließt — the relay closes
4 das Relais öffnet — the relay opens
5 das Relais ist offen — the relay is open
6 das Relais bleibt ⌈erregt [angezogen]; das Relais hält sich — the relay remains energized
7 bei angezogenem Relais K8 — with K8 energized
8 die Relais K2 und K3 ziehen an — the K2 and K3 relays pick up; the relays K2 and K3 pick up
9 K2 und K3 öffnen — the K2 and K3 open
10 das Relais überwacht die Versorgungsleitung — the relay monitors the supply line
11 das Relais spricht auf eine bestimmte Änderung des Stromes an — the relay responds to a certain change of current
12 das Relais E arbeitet in Abhängigkeit vom Relais D — the E relay operates depending on the D relay
13 der Stromkreis für das Relais K7 wird geschlossen — the circuit to the K7 relay is ⌈completed [established]

14 der Stromkreis für das Relais K7 wird über Minus – K7 – b – Erde geschlossen; der Stromkreis für K7 ⌐bildet [ergibt] sich über Minus – K7 – b – Erde	the circuit to ⌐K7 [the K7 relay] is completed as follows: from —DC, K7 coil, b, to +DC
15 der Kontakt b schließt den Stromkreis für P	the contact b completes the circuit to P
16 der Haltestromkreis [Selbstschluß] des Relais P wird ⌐unterbrochen [aufgelöst]	the holding circuit to the P relay is broken

„Ablesung von Meßwerten"

1 die Spannungen sind gleich	the voltages are equal
2 der Stromkreis ist ⌐konstant [unveränderlich]	the current is constant
3 der Strom bleibt ⌐konstant [unveränderlich]	the current remains constant
4 die Spannung wird konstant gehalten	the voltage is held constant; the voltage is maintained constant; the voltage is left constant
5 die Werte ⌐schwanken [bewegen sich] zwischen 1 und fast 0	the values vary from unity to nearly zero
6 die Zahlen liegen zwischen 100 und 150	the figures lie between 100 and 150
7 der Widerstand ist klein	the resistance is ⌐low [small]
8 die Spannung ⌐nimmt ab [fällt ab; sinkt ab; sinkt; verringert sich; verkleinert sich; vermindert sich; reduziert sich; geht zurück]	the voltage ⌐decreases [is decreasing; falls]
9 der Überlastfaktor soll nicht weniger als 5 dB betragen	the overload factor shall be no less than 5 dB
10 die Spannung sinkt auf das 0,707fache des Mittelwertes	the voltage falls to 0.707 of the mean value
11 die Spannung sinkt um 10% ab	the voltage falls off by 10%
12 die Leistung ist ⌐groß [hoch] genug	the power is large enough
13 der Frequenzbereich ist groß	the frequency range is wide
14 die Spannung ⌐nimmt zu [steigt an; steigt; erhöht sich; wächst; wird größer]	the voltage ⌐increases [is increasing; grows; rises; is rising; picks up]
15 den Nennwert um 10% überschreiten	to exceed the rated value by 10%
16 der Wert strebt nach unendlich	the value approaches infinity
17 der Leistungsfaktor ist 95% oder besser	the power factor is 95% or better
18 den Frequenzgang verbessern	to improve the frequency response

Sachregister

der in den „Fachsprachlichen Anmerkungen" kommentierten englischen Fachbenennungen.

Bemerkung: Die Stichwörter in diesem Sachregister sind nach der im Englischen als „reading-right-through" bezeichneten Art geordnet, zum Unterschied von der sog. „word-by-word"-Ordnung, die im Deutschen üblich ist.

A

a. c. component 7
accumulator 53
acoustic(al) 8
a.c. switch 92
active sensor 46
aerial 198
aerial cable 198
aerial line 198
a-f 169
airborne 240
alive 36
allocation (frequency) 181
allotment (frequency) 181
alphabetic(al) 8
alternating component 7
ALU 139
AM broadcasting 182
amplitude 7
amps 44
annotation (comp.) 139
antenna 198
antenna polar diagram 206
antinode 45
aperture 198
aperture efficiency 198
aperture illumination 198
aperture radiator 198
apex 199
apex driven antenna 199
arithmetic (and logical) unit 139
array 199
art 35
artificial mains network 27
A-scope 230
assignment (frequency) 181
atmospheric duct 79
attenuation 45
attenuator 147
audio 169
audio engineering 169
audio frequency range 169
auditory 169
aural 169
aural radio range 169

aural signal 169
aural transmitter 169
automatic closed loop control 129
automatic production 128
automation 128
automation methods 129
autosyn 66
auxiliary operation 139
availability 35
avalanche diode 93
average pulse spacing 115
AWG 28

B

backfire antenna 199
backswing 114
balance; balancing 158
balanced-unbalanced 158
balun 158
barrier layer 91
basic Q 104
battery 53
beacon 241, 242
beating 191
bend 217
(to) be on the air 182
bidirectional switch 92
binary system 141
black level 230
blanking level 230
blanking signal 230
blip 240
blocking layer 92
break, break 159
break, break, make 159
breakdown diode 93
break, make SPDT 159
break SPSTNC 159
bridge balancing 158
broadcasting 181
BSWG; B&SWG 28
(to) build up 26
built-up material 26
built-up mica 26
bunching 215
buried switch 59
buzz 192

BWG 28

C

cable television 228
capability 35
capacitance 8
capacitor 8, 35
capacitor balancing 158
capacitor microphone 170
capacity 8
cascade image tube 104
CATV 228
CCTV 228
cell (galvanic) 53
Celsius degree 19
centigrade 19
central processing unit 138
central processor 138
Certification Scheme 36
challenger 243
chapter 74
character 148
checking 129
circuit 130
circular mil 45
clause 74
clerical operation 139
click 192
closed circuit television 228
closed loop control 129
cmil 45
coaxial line 216
coefficient of self-induction 9
collating 139
collector 92
collinear array 199
columbium 105
combinational logic circuit 140
command signal 129, 130
community antenna television 228
complex sound 170
compliance 169
composite 230
composite material 26

281

composite signal 169
compound action 129
compound signal 169
Compton effect 74
computer instruction code 140
computing speed 66
condenser 35
condenser bushing 35
condenser microphone 35, 170
conductance 9
conductivity 9
conductor 9
conical horn 215
connector 139
contact functions 158
continuous action 129
continuous component 7
continuous noise 192
control 129
control action 129
control circuit 130
controlled switch 92
controlled variable 130
control unit 138
controlware 140
conventional 74
converted command signal 129
converter 53, 191
converter tube 104, 191
convertor 53
cordless 192
corner 217
corner aerial; corner antenna 241
corner reflector 241
counterpoise 199
counterweight 199
(to) cover 182
coverage 182
coverage diagram 182
coverage distance 182
coverage radius 182
(to) cover an event 182
CPU 138
crest value 7
current-carrying 36
customary temperature 20
cycle 44

D

db; dB 205
dba 206
dbk 206
dbm 206
dbrap 206
dbrn 206
dbv 206
dbw 206
d. c. component 7
DDC 130
decibel 205
decilog 205
decimal system 141
decision 139
degree 19
degree (Celsius, Kelvin) 19
dependability 35
depletion layer 91
detecting element 45
deviation 130
diac 92
diagram 36
diffused alloy transistor 91
digital control 130
digital integrated circuit 92
digital voltmeter 130
dihedral corner reflector 241
direct digital control 130
directional pattern 206
direct wave 206
discontinuous action 129
dislocation 91
distance 242
distortion 148
distortion factor 8, 148
document 139
doping 91
double amplitude 7
double-break (contacts) 158
double image 229
double-make (contacts) 158
double-pole (switch) 158
double-throw (switch) 158
doughnut 65
downcoming wave 207
down-converter 191
drain 92
drift transistor 91
dubbing 171
duodecimal system 141
durchgriff 104
duty factor 115
dynamic loudspeaker 170
dynamic microphone 170
dynistor 92

E

earth 148
earthed 148
earth-guided wave 206
earth satellite 148
earth station 148
echoes 241
echo signal 229
effect 74
effective radiated power 199
effective sound pressure 169
EIRP 199
elbow 217
electric(al) 7
electrical art 35
electrical energy 8
electrical engineer(ing) 8, 27
electrical insulator 8
electrical machinery 8
electrical megaphone 170
electrical unit 8
electric cable 8
electric field 8
electric flux 20
electric generator 8
electric pad 148
electric tension 20
electroacoustics 169
electrodynamic loudspeaker 170
electrodynamic microphone 170
electrogalvanizing 79
electromagnetic interference 191
electromagnetic loudspeaker 170
electromagnetic microphone 170
electromagnetic moment 20
electromagnetic noise 192
electronically 8
electronic engineer 8
electronic megaphone 170
electrostatic microphone 170
electrotechnical 8, 27
electrotechnique 27
emitter 92
emitter junction 92
E mode 216
endfire antenna 199
energy 53

engineering 27
E-plane sectoral horn 215
equivalent isotropically radiated power 199
(to) erase 59
ERP 199
error 130
error signal 130
excited-field loudspeaker 170
external Q 104

F

family 103
family of curves 103
fan (marker) beacon 242
far field 199
fast film 66
fast lens 110
f.h.p. motor 65
field 228
field blanking 229
field effect 74
field frequency 229
field hold control 60
field pattern 206
filament 104
filamentary cathode 104
filamentary transistor 104
firmware 140
fixed attenuator 147
flare 215
flowcharting symbols 139
flow-line 139
fluorescence 27
flush mounting 59
flux of displacement 20
f-number 110
four-course beacon 242
four-layer diode 92
four-pole network 148
four-terminal network 148
fractional horse power motor 65
frame 228
Fraunhofer diffraction region 199
frequency 35
frequency-change oscillator 191
frequency changer 53, 191
frequency converter; frequency convertor 53, 191
frequency distribution 35

frequency modulation 190
frequency range 242
fuel battery 53
fuel cell 53
functional design 140

G

gain control 129
galvanic cell 53
galvanism 79
galvanization 79
galvanizing 79
gate 92
general purpose television 228
(to) generate 139
ghost 229, 241
ghost echo 241
ghost image 229, 241
ghost mode 229
ghost pulse 241
ghost signal 241
G.P. television 228
grade 19
graded-base transistor 91
ground 148
ground clutter 148
grounded 148
ground-reflected wave 206
ground return 148
ground return circuit 148, 158
ground state 148
ground wave 206
group velocity 66
guide wire 216

H

halflife 200
half-power beam width 199
half-value 199
half-value layer 200
hardware 140
harmonic component 8
harmonic content 8
harmonics 8
helical 74
hexadenary system 141
HF-broadcasting 182
hi-fi 169
high fidelity 169
high-tension battery 80
H mode 216
hoghorn 215
hold 59
hold range 60

hook-collector transistor 92
hook transistor 92
hop 206
horn 215
horn antenna 215
hot 36
hot galvanizing 79
hot laboratory 36
hot line 36
hot reserve 36
H-plane sectoral horn 215
H.-T. battery 80
H.T. cable 80
H.T. winding 80
hysteresis motor 65

I

IEC 8, 27
i.h.p. motor 65
image 229
image antenna 229
image constant 229
image contrast 229
image converter tube 104
image frequency 229
image iconoscope 229
image impedance 229
image intensifier tube 104
image signal 229
image tube 104, 229
imperfection 91
impulse 44
impulse excitation 44
impulse noise 192
impurity 91
inductance 9
induction motor 65
inductive reactance 9
inductor 9
information 148
information circuit 130
information storage and retrieval 141
information theory 148
infrared maser 109
infrasonic frequency 171
input 54
input/output 139
inspection 129
instant replay 171
instruction code 140
(to) insulate 26
insulated supply system 27
insulation 26
insulator 26

integral horse power motor 65
intelligence 148
interference 170, 191
international broadcasting 182
interrogator 243
interrogator-responsor 243
interstitial atoms 91
intrinsic Q 104
inverted rotary convertor 53
inverter 53
ionospheric wave 207
IR (inform.) 141
(to) irase 109
iraser 109
(to) isolate 27
isolated neutral system 27
isolating network 27
isolating unit 27
isolation 26
isolation network 27
isolations switch 27
isolator 27
isolator switch 27
ISR (inform.) 141

J

jamming 191
jeep television 228
jump 206
junction 91
junction circuit 159
junction transistor 91

K

kc 44
kelvin 19
kilo 19
kilogramme-force 19
kilomegacycle 19
kmc 19

L

land 147
(to) lase 109
laser 109
laser pickup 109
lattice vacancy 91
lbf 19
lead 27
leader 27
lead-in groove 27
lens 109
lens speed 110
lens stop 110

level 110
lifeware 140
lift 230
linear integrated circuit 92
line balancing 158
line-of-sight coverage 182
live 36
live load 36
live steam 36
live transmission 36
loaded Q 104
local oscillator 191
locked 66
lock-in 59
locking 60
lock-in range 60
logic(al) 140
logic(al) design 140
logical operation 140
logic element 140
logic function 140
logic variable 140
long backfire 199
loop 45
loud hailer 170
loudness 170
loudness level 170
loudspeaker 170, 171
low tension 80

M

machine (instruction) code 140
machine language 140
machine readable form 140
machine script 140
magnetic armature loudspeaker 170
magnetic biasing 169
magnetic loudspeaker 170
magnetic microphone 170
magnetic moment 20
magslip 66
maintainability 35
make, break, continuity transfer 159
make, make 159
make SPSTNO (switch) 159
man-made noise 192
manual operation 139
marker 241
marker beacon 241
marker pip 241
maser 109

maximum speed 66
mc 44
measured value of the controlled condition 130
measuring element 45
measuring technique 27
measuring transmitter 45
measuring unit 45, 46
mechanical 8
medium-wave broadcasting 182
mega 19
megacycle 44
megaphone 170
memory 141
mess 66
message 148
mess-motor 65
metallic circuit 158
metrication 45
mho 20, 44
micanite 26
micro 19
microphone 170
microwave range 243
mike 170
mil 28, 45
mil-foot 45
mill 45
mixer 191
mode 116, 215
modulated 79
modulation 79
moment 20
moment of a couple 20
moment of momentum 20
momentum 20
monitor 230
monitored control 129
motor convertor 53
moving armature loudspeaker 170
moving coil loudspeaker 170
moving-conductor loudspeaker 170
moving-conductor microphone 170
moving-iron loudspeaker 170
moving iron microphone 170
moving target indicator 241
MTI radar 241
multihop 206
multiple image 229

mutual conductance 20, 104

N

natural noise 192
NBS 28
nc 130
NC (switch) 158
needle (phonograph) 171
negative-resistance switch 92
nemo 182
network balancing 158
noise 170, 191
noise ratio 191, 192
non-directional (antenna) 182
non-erasable store 141
non-loaded Q 104
normally closed (switch) 158
normally open (switch) 158
notation 140
not emanating from main office 182
number of revolutions 20
numerical control 130
numerical machine tool control 130

O

O.B. 182
objective 109
O.B. van 182
octal system 141
octonary system 141
off-line store 139
omnibearing (antenna) 182
omnibearing range 183
omnidirectional (antenna) 182
omnidirectional range 183
omnirange 183
on-line store 139
on the air 182
opaque 110
open loop control 129
optical 8
optical maser 109
orbit 110
orbital 110
orthodox 74
output 54
outside broadcasting 182
overshoot 114
overspeed 66

oxidation 79
oxidizing 79

P

pad 147
padder 148
padding 148
pad resistance 147
pad roll 147
pair 217
PAM 115, 191
panel 59
paragraph 74
part 74
passive sensor 46
(to) patch 59
path 110
PCM 115, 191
PCM/PM 115
PDM 115
peak-to-peak value 7
peak value 7
pedestal 230
penetration factor 104
per cent ripple 9
period 20, 79
periodic 79
periodic pulse train 116
periodic time 20
permanent magnet motor 65
permanent store 141
PFM 115
phase constant 45
phase convertor 53
phase lead 27
phase modulation 190
phenomenon 74
phosphor 27
phosphorescence 27
phosphorus 27
pickup 45
pico 19
picture 228
picture monitor 230
picture signal 230
ping 240
pip 240
plan 36
plane antenna 199
plane reflector 199
playback 171
playback head 171
play-over 171
PM 190
p-n-p-n switch 92
point contact transistor 92
polar diagram 206

poles (switch) 159
position notation 140
positive feedback 66
potential difference 80
potential jump 206
poundal 19
pound-force 19
power 53
power circuit 130
power engineering 54
power equipment 54
power industry 54
power pack 54
power series 54
power stage 54
power tool 54
PPM 115
predefined process 139
PRF 115
primary cell 53
primary detector 45
primary element 45
probability 35
processing 139
program flowcharting symbols 139
program modification 139
progressive action 129
project 36
propagation constant 45
PRR 115
PTM 115
pull(ing)-in 59
pulsating quantity 7
pulse 44
pulse amplitude modulation 115, 191
pulse-code modulation 115, 191
pulse-count modulation 115
pulse duration (modulation) 115
pulse duty factor 115
pulse-frequency modulation 115
pulse group 116
pulse interval 115
pulse length (modulation) 115
pulse mode 116
pulse modulation 115, 190
pulse-phase modulation 115
pulse-position modulation 115
pulse rate 116

285

pulse recurrence frequency 115
pulse recurrence rate 116
pulse repetition frequency 115
pulse repetition period 115
pulse repetition rate 115
pulse repetition rate modulation 115
pulse separation 115
pulse spacing 115
pulse-time modulation 115
pulse train 116
pulse width (modulation) 115
pure sound 172
(to) put on the air 182
pyramidal horn 215

Q

Q 104
Q external 104
Q-loaded 104
Q-non-loaded 104
quadripole network 148
quadrupole network 148

R

racon 242, 243
radar 242, 243
radar beacon 242, 243
radar range 242
radar transponder 243
radiation diagram 206
radiation pattern 206
radio 192
radio beacon 242
radio detection and ranging 242
radio direction and range 242
radio-frequency isolating filter 27
radio interference 191
radio marker beacon 241
radio noise 192
radio range 242
radix 141
radix notation 140
RAM 141
random access memory 141
random noise 192
range 182, 242
range-height indicator 242

reactance 9
reactor 9
read only memory 141
recessed 59
reciprocal amplification factor 104
reconstructed 26
reconstructed mica 26
record signal 169
recovered 26, 66
recovered mica 66
recuperative 66
rediffusing broadcasting 182
rediffusion 182
reference input 129
reference quantity 129
reference variable 129
regenerated oil 66
regenerative 66
regenerative coupling 66
regulated power supply 130
regulation 130
regulator 93, 130
relative harmonic component 8
relay contact functions 158
reliability 35
remica 26, 66
remote transmission 182
repeatability 35
replay 171
reproduction 171
reproduction of a recording 171
re-recording 171
resistance 8
resistance temperature coefficient 21
resistivity 9
resistor 8
resolving power 54
responder 243
responser 243
restorability 35
retrieval 141
(to) retrieve 141
revolutions per second 20
ripple 9
ripple quantity 9
ripple ratio 9
ROM 141
rotary convertor 53
rotational frequency 20
rotational speed 20, 66
r.p.m. 20
r.p.s. 20

rules of the art 35

S

schedule 36
scheme 36
Science and Arts 35
SCR 92
secondary cell 53
section 74
sectoral horn 215
self-acting 75
self-contained 74
self-healing 75
self-inductance 75
self-operated sensors 46
self-synchronous 66
selsyn 66
sensing element 45
sensor 45
sequence 116
service diagram 182
set of symbols 148
set point 129
set-up 230
set value 129
sexadenary system 141
short backfire 199
sidac 92
signal converter 46
signal-to-noise ratio 192
silent zone 206
silica 36
silicon 36
silicon controlled rectifier 92
silicone 36
silicone oil 36
silicon rectifier 36
silicon steel 36
single hop 206
single-pole 158
single-throw 158
single-wire transmission 216
skating 172
skip 206
skip area 206
skip distance 206
skip zone 206
sky wave 206
SNR 192
S/N ratio 192
software 140
solar battery 53
solar cell 53
solid state image converter 104
solid state image intensifier 104

solid state thyratron 92
sorting 139
sound 169
sound broadcasting 182
sound intensity 169
sound level 169
source 92
space charge layer 92
space wave 206
spacing 115
spacing interval 115
SPDT (switch) 159
speaker 171
speech frequency 169
speed 66
speed limit 66
speed of rotation 20
speedometer 66
spiral 74
SPK'R 171
split-field motor 65
SPSTNC (switch) 159
square mil 45
stalled 66
standard broadcasting 182
standards converter 230
star connection 60
state of the art 35
static condenser 35
step-by-step action 129
step-by-step control 129
stilus 171
storage 141
storage battery 53
storage cell 53
store 139, 141
stylus 171
stylus force 171
stylus pressure 171
sub-audio frequency 171
sub-audio telegraphy 171
substitutional atoms 91
sunk mounting 59
super-audio frequency 171
supersonic 171
supersonic aircraft 171
surface 79
surface duct 79
surface mounting 59
surface-to-air missile 79
surface vessel 79
surface warning radar 79
surface wave 206
surveillance 129
symbol 148
sync 59
synchro 60, 66

synchronism 60
synchronization 60
synchronizing signal 230
synchronous 60, 66
synchronous condenser 35
synchronous motor 65
synchro transformer 66
sync level 60
system flowcharting symbols 139

T

T.C. 20
technique 27
technological gap 27
technological progress 27
technology 27
telecasting 182
television broadcasting 182
television frequency converter 230
TEM mode 216
TE mode 216
temperature 19, 20
temperature coefficient 20
tension 20, 80
terminal 139
terminal area 147
ternary system 141
testing 129
thermodynamic temperature 19
thou 45
three-level laser 110
three-term action 130
throughput 54
throws (switch) 159
thyristor 92
TM mode 216
toll circuit 159
tone 171
tone control 172
tonf 19
ton-force 19
torque 20
torque motor 65, 67
torque synchro 67
torquer 67
total amplitude 7
trace 79
tracing distortion 172
track 79
trackability 172
tracking error 172
trajectory 110
transconductance 20, 104

transducer 45
transfer SPDT (switch) 159
(to) transistorize 92
translator 230
translucent 110
transmission mode 215
transmitter responder 243
transparent 110
transponder 243
transposer 230
trap 91
trapping centre 91
triac 92
trihedral corner reflector 241
trinistor 92
trunk circuit 159
TTL, T^{2}L 141
tungstate 105
tungsten 105
turn 67
TV frequency converter 230
twin lead 216
twin wire 216
two-level action 129
two-port network 148
two-position action 129
two-step action 129
two-term action 130
two-terminal-pair network 148
two-wire line 216

U

ultrasonic frequency 171
umho 20
undershoot 114
ungrounded system 27
unloaded Q 104
unmonitored control 129
up-converter 191
uu; UU 19

V

vacuum-tube family 103
variable reluctance microphone 170
velocity 66
velocity modulation 66
vicenary system 141
video 230
vision 230
visual 230
voice 169
voice frequency 169
voltage 20, 80

voltage reference diode 93
voltage regulator diode 93
voltage standing-wave ratio 9
volume 170
volume control 170
volume expansion 170
V reflector aerial 241
VSWR 9
vu-units 170

W

wave 206
waveform monitor 230
waveguide 216
waveguide bend 217
waveguide corner 217
waveguide elbow 217
wheel balancing 158
winding 67
wire broadcasting 182
wired broadcast 182
wired radio 182
wire ga(u)ge 28
wireless 192
wireless microphone 192
wolfram 105
wolframate 105
wolframite 105
wolframium 105
working Q 104
wye connection 60
wye-grounded 60

Y

Y-connection 60
yearly output 54

Z

Z-diode 93
zener diode 93
zero crossings per second 116
zeroes per second 116
zps 116